T0135050

# Smart Innovation, Systems and Technologies

## Volume 88

**Series editors**

Robert James Howlett, Bournemouth University and KES International,
Shoreham-by-sea, UK
e-mail: rjhowlett@kesinternational.org

Lakhmi C. Jain, University of Canberra, Canberra, Australia;
Bournemouth University, UK;
KES International, UK
e-mails: jainlc2002@yahoo.co.uk; Lakhmi.Jain@canberra.edu.au

The Smart Innovation, Systems and Technologies book series encompasses the topics of knowledge, intelligence, innovation and sustainability. The aim of the series is to make available a platform for the publication of books on all aspects of single and multi-disciplinary research on these themes in order to make the latest results available in a readily-accessible form. Volumes on interdisciplinary research combining two or more of these areas is particularly sought.

The series covers systems and paradigms that employ knowledge and intelligence in a broad sense. Its scope is systems having embedded knowledge and intelligence, which may be applied to the solution of world problems in industry, the environment and the community. It also focusses on the knowledge-transfer methodologies and innovation strategies employed to make this happen effectively. The combination of intelligent systems tools and a broad range of applications introduces a need for a synergy of disciplines from science, technology, business and the humanities. The series will include conference proceedings, edited collections, monographs, handbooks, reference books, and other relevant types of book in areas of science and technology where smart systems and technologies can offer innovative solutions.

High quality content is an essential feature for all book proposals accepted for the series. It is expected that editors of all accepted volumes will ensure that contributions are subjected to an appropriate level of reviewing process and adhere to KES quality principles.

More information about this series at http://www.springer.com/series/8767

Sajal K. Das · Nabendu Chaki

Editors

# Algorithms and Applications

ALAP 2018

 Springer

*Editors*
Sajal K. Das
Department of Computer Science
Missouri University of Science
  and Technology
Rolla, MO
USA

Nabendu Chaki
Department of Computer Science
  and Engineering
University of Calcutta
Kolkata, West Bengal
India

ISSN 2190-3018             ISSN 2190-3026  (electronic)
Smart Innovation, Systems and Technologies
ISBN 978-981-13-4065-9      ISBN 978-981-10-8102-6  (eBook)
https://doi.org/10.1007/978-981-10-8102-6

Printed on acid-free paper

This Springer imprint is published by the registered company Springer Nature Singapore Pte Ltd. part of Springer Nature
The registered company address is: 152 Beach Road, #21-01/04 Gateway East, Singapore 189721, Singapore

# Preface

The first International Conference on Algorithms and Applications (ALAP 2018) was held on January 10–12, 2018, at B. P. Poddar Institute of Management and Technology (BPPIMT), Kolkata, India. The technical program was organized in three focus areas or tracks—VLSI and Embedded Systems, Distributed Systems and Security, and Big Data and Analytics. The tracks were chaired by Prof. Kolin Paul (IIT Delhi) and Prof. Kamalika Datta (NIT Meghalaya) for VLSI and Embedded Systems; Prof. Pradip K. Das (formerly Jadavpur University) for Distributed Systems and Security; and Dr. Arijit Mukherjee (TCS Research) for Big Data and Analytics.

Thanks to the sincere effort of the conference stakeholders, in the very first year of its existence, ALAP 2018 attracted a modest 36 submissions. A thorough peer-review process carried out by the program committee members and reviewers mainly looked at the novelty of the contributions, besides the technical content and clarity of the presentation. The Technical Program Committee eventually identified only 6 papers for publication out of 36 submissions, resulting in the competitive acceptance ratio of 16.6%. The technical program also included several keynote and invited talks by eminent speakers from industry and academia, and pre-conference tutorials by different research groups at TCS Innovation and CDAC–Kolkata.

The overall quality of ALAP 2018 marks the success of the conference in bringing excellent researchers working in the broad domain of Algorithms and Applications into a common forum for exchanging ideas toward achieving greater scientific goals.

The conference would not have been successful without tremendous help and support of many individuals. We extend our heartfelt gratitude to the Track Chairs for their leadership in making a successful technical program. We also thank the members of the Program Committee for their excellent and time-bound review work. We are grateful to the management of BPPIMT for their warm patronage and support to make the event a great success. Our special thanks to Prof. Sutapa Mukherjee (Principal, BPPIMT) for her inspiring presence throughout the event, and Prof. B. N. Chatterji (Dean, BPPIMT) and Prof. Indranil Sengupta (IIT Kharagpur) for their critical roles as not only General Chairs of ALAP 2018, but also toward conceptualizing the conference and helping us since the first meeting held in early 2017. We appreciate the role of Prof. Ananya Kanjilal as Organizing Chair for ALAP 2018. She and her colleagues took care of every detail that ensured the success of the conference.

We appreciate the initiative and support of Mr. Aninda Bose and his colleagues at Springer Nature for their strong support toward publishing this volume. Finally, we thank all the authors and participants without whose presence and support the conference would not have reached the expected standards.

Rolla, USA                                                                      Sajal K. Das
Kolkata, India                                                           Nabendu Chaki
April 3, 2018

# Committee Members

## Chief Patron

Shri Arun Poddar, Chairman, B. P. Poddar Group

## Patrons

Shri Ayush Poddar, Vice Chairman, B. P. P. Institute of Management and Technology (BPPIMT)
Dr. Subir Choudhury, Founder Trustee & Chief Mentor, BPPIMT
Prof. (Dr.) Sutapa Mukherjee, Principal, BPPIMT

## General Chairs

B. N. Chatterji, BPPIMT, Kolkata, India
Indranil Sengupta, IIT Kharagpur, India

## Program Chairs

Sajal K. Das, Missouri University of Science and Technology, USA
Nabendu Chaki, University of Calcutta, India

## Organizing Chair

Ananya Kanjilal, BPPIMT, Kolkata, India

## Track Chair for Distributed Systems and Security

Pradip K. Das, Jadavpur University (Retd.), India

## Track Chair for VLSI and Embedded Systems

Kolin Paul, IIT Delhi, India
Kamalika Datta, NIT Meghalaya, India

## Track Chair for Big Data and Analytics

Arijit Mukherjee, TCS Research, India

## Program Committee

Aditya Bagchi, Ramakrishna Mission Vivekananda University, India
Agostino Cortesi, Ca Foscari University, Italy
Ajay K. Datta, University of Nevada, USA
Amlan RayChaudhuri, BPPIMT, Kolkata, India
Animesh Dutta, NIT Durgapur, India
Anupam Chattopadhyay, NTU, Singapore
Arnab Sarkar, IIT Guwahati, India
Avik Ghose, TCS Innovation and Research
Bhabani. P. Sinha, ISI Kolkata (retired), India
Bhargab B. Bhattacharya, ISI Kolkata, India
Bikromadittya Mondal, BPPIMT, Kolkata, India
Debatosh Bhattacharya, Jadavpur University, India
Deepak Kumar, NIT Meghalaya, India
Dhruba Kr Bhattacharyya, Tezpur University, Assam, India
Diptendu Sinha Roy, NIT Meghalaya, India
Esref Adali, Istanbul Technical University, Turkey

Hayato Yamana, Waseda University, Japan
Jayeeta Chanda, BPPIMT, Kolkata, India
Jimson Mathew, IIT Patna, India
Jiannong Cao, The Hongkong Polytechnic University, Hongkong
Jyoti Prakash Singh, NIT Patna, India
K Chandrasekaran, NIT Suratkal, India
Khalid Saeed, Bialystok University, Bialystok, Poland
Mohammad Sarosh Umar, Aligarh Muslim University, India
Manoj Singh Gaur, MNIT, Jaipur, India
Nabanita Das, ISI Kolkata, India
P. Rangababu, NIT Meghalaya, India
Pabitra Mitra, IIT Kharagpur, India
Paolo Bellavista, University of Bologna, Italy
Piotr Porwick, University of Silesia, Poland
Rajarshi Ray, NIT Meghalaya, India
Raju Halder, IIT Patna, India
Rituparna Chaki, University of Calcutta, Kolkata, India
Robert Wille, Johannes Kepler University Linz, Austria
Sabnam Sengupta, BPPIMT, Kolkata, India
Samiran Chattopadhyay, Jadavpur University
Sankhayan Choudhury, University of Calcutta, Kolkata, India
Santanu Chattopadhay, IIT Kharagpur, India
Sanjay Saha, Jadavpur University, India
Sarmistha Neogy, Jadavpur University
Shameek Bhattacharjee, Missouri University of Science & Technology, Rolla, USA
Shikharesh Majumdar, Carleton University, Canada
Soumya Sen, University of Calcutta, India
Sounak Dey, TCS, India
Supratik Chakraborty, IIT Bombay, India
Tapas Chakravarty, TCS Research, India
Vaskar Roychoudhury, IIT Roorkee, India

# Contents

# About the Editors

**Sajal K. Das** is a Professor of Computer Science and Daniel St. Clair Endowed Chair at Missouri University of Science and Technology, USA, where he was Computer Science Department Chair from 2013 to 2017. Prior to 2013, he was a University Distinguished Scholar Professor of Computer Science and Engineering at UT Arlington, USA. He also served as an NSF Program Director during 2008–2011. His research interests include IoT, big data analytics, security, cloud computing, wireless sensor networks, mobile and pervasive computing, cyber-physical systems and smart environments, biological and social networks, and applied graph theory. He has directed high-profile, funded projects and published over 700 papers in journals and conference proceedings. He holds 5 US patents and has co-authored 52 chapters and 4 books. A recipient of 10 best paper awards, he has also received numerous awards for teaching, mentoring, and research including IEEE Computer Society Technical Achievement Award for pioneering contributions to sensor networks and mobile computing. He is the founding editor-in-chief of the Pervasive and Mobile Computing Journal and associate editor of several other journals. He is an IEEE fellow.

**Nabendu Chaki** is a Professor at the Department of Computer Science and Engineering, University of Calcutta, Kolkata, India. He shares seven international patents including four US patents. In addition to nearly 30 edited volumes with Springer, he has authored 7 text and research books and more than 150 Scopus Indexed research papers in journals and at international conferences. He has been a visiting professor at various institutions, including Naval Postgraduate School, USA; Ca' Foscari University, Italy; and AGH University of Science and Technology, Poland. He is the Founder Chair of the ACM Professional Chapter in Kolkata and served in that capacity for 3 years from January 2014. During 2009–2015, he was active in developing several international standards in software engineering and service science as a global (GD) member of ISO-IEC.

# Part I
# VLSI and Embedded Systems

# Taxonomy of Decimal Multiplier Research

**Diganta Sengupta** and **Mahamuda Sultana**

**Abstract** Decimal arithmetic hardware research accelerated in the last decade with introduction of decimal floating point formats in "IEEE 754-2008" standards. During the revision phase (IEEE 754R), 2000–2008, global research on decimal arithmetic witnessed state-of-the-art decimal hardware proposals as well as software routines for decimal computations on general-purpose microprocessors. Multiplication forms a fundamental arithmetic operation and an integral part of arithmetic hardware units. This paper provides taxonomy of the major contributions in decimal multiplier research.

**Keywords** Decimal architecture · Fixed point multiplier
Floating point multiplier · Decimal floating point · Taxonomy

## 1 Introduction

With the advent of digital computers, researchers and computer scientists indulged in the worldwide debate whether the base architecture should be binary or decimal. Binary got the lead in 1946 with Burks, Goldstine, and Von Neumann advocating for binary architecture. On a different note, many researchers were of the view that the architecture should comprise of binary addressing supported by decimal data arithmetic [1]. Richards [2] provided a few decimal arithmetic architecture proposals as well as different decimal digit representation codes (i.e., 4221, 5211, etc.)

D. Sengupta (✉)
Techno India-Batanagar, Kolkata, India
e-mail: sg.diganta@gmail.com

M. Sultana
Techno India College of Technology, Kolkata, India
e-mail: sg.mahamuda@gmail.com

© Springer Nature Singapore Pte Ltd. 2018
S. K. Das and N. Chaki (eds.), *Algorithms and Applications*, Smart Innovation,
Systems and Technologies 88, https://doi.org/10.1007/978-981-10-8102-6_1

which initiated appreciable research [3, 4] in the previous decade for search of decimal multiplier hardware. Decimal arithmetic was primarily sustained by software implementations [5] and libraries [6–8] but at the expense of slower processing as performance penalty of software over hardware implementations is 100–1000 times. Intel was the first microprocessor manufacturer to introduce the 8087 numeric extension processor supporting 18 decimal digits along with additional software routines in 1983 [9] allowing easy compatibility with IEEE 754–1985. The 8087 coprocessor served the financial community and the Certified Public Accountants (CPAs) for almost 25 years. IEEE rolled out the "IEEE 754-2008" standards in 2008 defining two Decimal Floating Point (DFP) formats, decimal 64 and decimal 128, having a precision of 16 and 34 digits, respectively. With continual depreciation of die space expense and potential speedup achievable in hardware implementations [10, 11], microprocessor manufacturers have already mirrored the current trend of decimal arithmetic hardware research by commercializing digital processors with embedded decimal arithmetic units [12–14]. DFP multiplier architecture proposals comprise of decoding of the *Densely Packed Decimal* (*DPD*) numbers [15] into equivalent BCD format, decimal multiplication, cohort selection [16], rounding [16–22], and eventual encoding [15] in the DFP format with a few other important intermediate steps.

We provide taxonomy of decimal multiplier hardware designs in this paper. We have generated the vocabulary from *IEEE Xplore* using "decimal multiplication/multiplier" as the *seed term*. The taxonomy has been created manually using steps mentioned in *Automated Taxonomy Generation Process* based on a bibliometric method presented in [23] keeping the *seed term* constant.

## 2 Article Classification

A broad collection of research articles form the prerequisites for conducting a survey as well as generating taxonomy. In purview of taxonomy, raw data refers to the published research articles. We have identified "*IEEE Xplore*" for generating the vocabulary of raw data. "Decimal multiplication/multiplier" was used as the seed term to shortlist the relevant articles from the online database. Each article in the raw data was allotted a "Document Id" for further processing. Table 1 provides the necessary allotments for all the published articles till date. The time span for the survey was divided into three parts depending upon IEEE floating point standardizations:

            IEEE 754 - 985 period       : Before year 2000 : Period 1
            IEEE 754R revision period : Year 2000−2008 : Period 2
            IEEE 754 - 2008 period     : Year 2000−2016 : Period 3

**Table 1** Document Id assignment for published literature

| DiD | Ref. | DiD | Ref. | DiD | Ref. | DiD | Ref. | DiD | Ref. | DiD | Ref. |
|---|---|---|---|---|---|---|---|---|---|---|---|
| PL1 | [2] | PL2 | [4] | PL3 | [24] | PL4 | [25] | PL5 | [26] | PL6 | [18] |
| PL7 | [27] | PL8 | [28] | PL9 | [29] | PL10 | [30] | PL11 | [31] | PL12 | [5] |
| PL13 | [32] | PL14 | [33] | PL15 | [34] | PL16 | [35] | PL17 | [36] | PL18 | [37] |
| PL19 | [38] | PL20 | [39] | PL21 | [40] | PL22 | [41] | PL23 | [42] | PL24 | [43] |
| PL25 | [44] | PL26 | [20] | PL27 | [45] | PL28 | [46] | PL29 | [47] | PL30 | [21] |
| PL31 | [48] | PL32 | [49] | PL33 | [50] | PL34 | [17] | PL35 | [13] | PL36 | [22] |
| PL37 | [51] | PL38 | [52] | PL39 | [53] | PL40 | [3] | PL41 | [54] | PL42 | [55] |
| PL43 | [56] | PL44 | [57] | PL45 | [58] | PL46 | [59] | PL47 | [60] | PL48 | [61] |
| PL49 | [62] | PL50 | [63] | PL51 | [64] | PL52 | [65] | PL53 | [66] | PL54 | [67] |
| PL55 | [68] | PL56 | [69] | PL57 | [70] | PL58 | [71] | PL59 | [19] | PL60 | [72] |
| PL61 | [73] | PL62 | [74] | PL63 | [75] | PL64 | [76] | PL65 | [77] | PL66 | [78] |
| PL67 | [79] | | | | | | | | | | |

DiD: Document Id; Ref.: Reference number; PL: Published Literature

Overall, 101 articles were extracted comprising of Journal as well Conference publications. 13 articles belonged to IEEE 754-1985 period and witnessed discrete publications, i.e., long gaps between publications. The rest 88 articles belonged post-2000 period. Of these 88 articles, 67 articles were found to be directly connected to decimal multiplier research and have been included in Table 1. Table 1 is mainly involved in generating the taxonomy.

## 2.1 Statistics of Decimal Multiplier Research

Research on decimal arithmetic majorly focused after the year 2000. Therefore, we provide a statistical analysis from year 2000 till date. Table 2 provides the year-wise article count for Journal and Conference publications. Article count prior to 2000 has been clubbed as a single entry as the number of published articles is highly discrete.

Figure 1 presents the publication count for Periods 1 through 3. If publication count can be assumed to resemble the quantum of global research, then it can be observed from Fig. 1 that global research on decimal multiplier hardware has been conducted after IEEE rolled out the IEEE 754-2008 standards [16].

Figure 2 provides the categorical (Journal/Conference) publication count using data from Table 2. It can be observed from Fig. 3 that the rate of publication through the years has been logarithmic in nature.

The following equations provide the rate of Journal and Conference publications, respectively, as observed from Fig. 3.

$$y = 0.579 \ln (x) + 0.681 \tag{1}$$

**Table 2** Year-wise article count for Journal and Conference publications

| Year | Publication type | Article count | Total | Year | Publication type | Article count | Total |
|---|---|---|---|---|---|---|---|
| <2000 | Journals | 13 | 13 | 2008 | Journals | 2 | 10 |
| | Conferences | 0 | | | Conferences | 8 | |
| 2000 | Journals | 2 | 2 | 2009 | Journals | 3 | 11 |
| | Conferences | 0 | | | Conferences | 8 | |
| 2001 | Journals | 0 | 2 | 2010 | Journals | 4 | 10 |
| | Conferences | 2 | | | Conferences | 6 | |
| 2002 | Journals | 2 | 3 | 2011 | Journals | 1 | 4 |
| | Conferences | 1 | | | Conferences | 3 | |
| 2003 | Journals | 0 | 2 | 2012 | Journals | 1 | 7 |
| | Conferences | 2 | | | Conferences | 6 | |
| 2004 | Journals | 1 | 2 | 2013 | Journals | 4 | 9 |
| | Conferences | 1 | | | Conferences | 5 | |
| 2005 | Journals | 1 | 3 | 2014 | Journals | 2 | 3 |
| | Conferences | 2 | | | Conferences | 1 | |
| 2006 | Journals | 1 | 2 | 2015 | Journals | 1 | 2 |
| | Conferences | 1 | | | Conferences | 1 | |
| 2007 | Journals | 3 | 12 | 2016 | Journals | 3 | 4 |
| | Conferences | 9 | | | Conferences | 1 | |

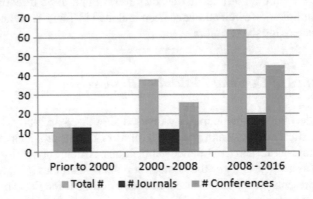

**Fig. 1** Total publication count

$$y = 1.983 \ln (x) + 0.268 \tag{2}$$

The equations reflect a growth in research in the previous years. One of the important measures of published materials is their citation count. The citation count can also be assumed to be a measure of research conducted on a certain design and the expansion and full exploration of the design with peers. Hence, we provide the citation count in Table 3 and Fig. 4 for all articles mentioned in Table 1. The median of

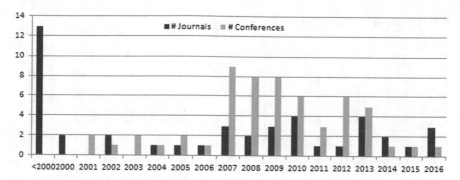

**Fig. 2** Year-wise categorical publication count

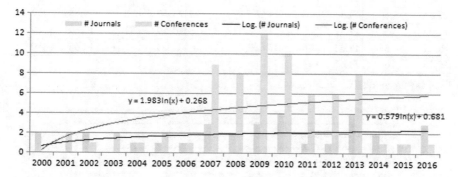

**Fig. 3** Mathematical expression for research on decimal multiplier research

the graph presented in Fig. 4 is 13, whereas the arithmetic, geometric, and harmonic mean values are 23.94, 11.95, and 5.49, respectively. Since the graph is populated with extreme outliers and no inter-related data, hence, we consider the harmonic mean value to be more accurate to give a measure of the citation trend. Also from the difference between the values of the arithmetic, geometric, and harmonic mean, we observe that the harmonic mean gives a more average count for citations. Therefore, going by the harmonic mean, we assume that research in any article that fetches a citation count greater than 5.49 has been further explored and expanded. Therefore, these can form the very pivotal publications in decimal multiplier research. Hence, we classify the articles into four classes based on their citation count as shown in Table 4.

## 3 Taxonomy Proposal

Taxonomy refers to the hierarchy of relevant terms in a particular research domain. Usually, it follows a top-down approach where the topmost node is a general research

**Table 3** Number of citations received by the respective articles

| DiD | CC | DiD | CC | DiD | CC | DiD | CC | DiD | CC |
|---|---|---|---|---|---|---|---|---|---|
| PL1 | 0 | PL2 | 24 | PL3 | 1 | PL4 | 8 | PL5 | 8 |
| PL6 | 121 | PL7 | 48 | PL8 | 0 | PL9 | 28 | PL10 | 3 |
| PL11 | 0 | PL12 | 67 | PL13 | 48 | PL14 | 31 | PL15 | 47 |
| PL16 | 0 | PL17 | 2 | PL18 | 1 | PL19 | 8 | PL20 | 25 |
| PL21 | 56 | PL22 | 48 | PL23 | 4 | PL24 | 29 | PL25 | 4 |
| PL26 | 25 | PL27 | 1 | PL28 | 62 | PL29 | 3 | PL30 | 18 |
| PL31 | 11 | PL32 | 4 | PL33 | 6 | PL34 | 30 | PL35 | 21 |
| PL36 | 19 | PL37 | 19 | PL38 | 16 | PL39 | 7 | PL40 | 63 |
| PL41 | 11 | PL42 | 9 | PL43 | 1 | PL44 | 17 | PL45 | 0 |
| PL46 | 0 | PL47 | 3 | PL48 | 3 | PL49 | 0 | PL50 | 0 |
| PL51 | 1 | PL52 | 6 | PL53 | 2 | PL54 | 8 | PL55 | 11 |
| PL56 | 8 | PL57 | 0 | PL58 | 4 | PL59 | 37 | PL60 | 82 |
| PL61 | 3 | PL62 | 26 | PL63 | 11 | PL64 | 7 | PL65 | 102 |
| PL66 | 16 | PL67 | 13 | | | | | | | |

DiD: Document Id; CC: Citation Count

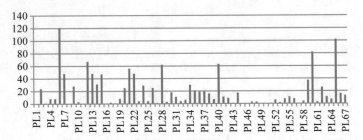

**Fig. 4** Citation counts for published literature

**Table 4** Citation count based on article classification

| Classification | Document Id |
|---|---|
| Greater than arithmetic mean | PL2, PL6, PL7, PL9, PL12, PL13, PL14, PL15, PL20, PL21, PL22, PL24, PL26, PL28, PL34, PL40, PL59, PL60, PL62, PL65 |
| Between arithmetic mean and median | PL30, PL35, PL36, PL37, PL38, PL44, PL66, PL67 |
| Between median and harmonic mean | PL4, PL5, PL19, PL31, PL33, PL39, PL41, PL42, PL52, PL54, PL55, PL56, PL63, PL64 |
| Below harmonic mean | PL3, PL10, PL17, PL18, PL23, PL25, PL27, PL29, PL32, PL43, PL47, PL48, PL51, PL53, PL58, PL61 |

term and the nodes at the bottom comprise of specific terms. The topmost node is generally referred to as the "root" and the nodes at the bottom as "leaves". The expansion from the single root to multiple leaves reflect the interrelationship between the concepts of research. There are many functional uses of taxonomy. We propose the taxonomy for better understanding of the research based on the characteristic features of "Decimal Multiplier Architectures".

## 3.1 Process of Taxonomy Generation

We have created the taxonomy using IEEE Xplore as the vocabulary and "Decimal Multiplier" as the seed term as mentioned earlier. The process described in the automated taxonomy generation methodology has been implemented manually comprising the following four steps.

Data Processing,
Database Creation,
Taxonomy Generation, and
Visualization.

The use of automated methodology warrants a certain degree of quality assurance for the results. For data processing, Table 5 was populated row-wise by relevant data from the database of articles in Table 1.

From Table 5, the keywords were extracted. We considered only those keywords whose document frequency was greater than 2, i.e., most frequently occurring keywords. The selected keywords/key phrases along with their respective weightages are presented in Table 6. The keywords/phrases are also termed as the nodal terms. These terms eventually form the nodes of the taxonomy.

We have used *Cosine Similarity* to detect the similarity between two terms. The angle between two terms is directly proportional to their similarity index, i.e., if the angle is "0", they are exactly identical and if the angle is 90°, they are 100% dissimilar as $Cos\,(0) = 1; Cos\,(90) = 0$. The similarity values range from 0 to 1 where 1 indicates exact similarity and 0 means independent (dissimilar) terms. The *Cosine Similarity* expression is provided in Eq. 3.

$$Cosine\ Similarity = \frac{n_{x,y}}{\sqrt{n_x}\sqrt{n_y}},\tag{3}$$

where

$n_{x,y}$ = Number of articles containing both term "x" and term "y",
$n_x$   = Number of articles containing only term "x", and
$n_y$   = Number of articles containing only term "y"

**Table 5** Data processing table

| DiD | Keywords | Abstract |
| --- | --- | --- |

**Table 6** *Term—Document* occurrence matrix

| Term Id | Nodal term/Phrase | Weightage | Term Id | Nodal term/Phrase | Weightage |
|---|---|---|---|---|---|
| TN1 | Decimal multiplication/Multiplier | 59 | TN10 | Computer arithmetic | 13 |
| TN2 | Parallel multiplication/Multiplier | 10 | TN11 | Decimal arithmetic | 29 |
| TN3 | Sequential multiplication/Multiplier | 5 | TN12 | Rounding | 19 |
| TN4 | Binary coded decimal/BCD | 35 | TN13 | Partial product generation | 11 |
| TN5 | Fused multiply–add | 3 | TN14 | Partial product reduction | 9 |
| TN6 | Decimal floating point | 30 | TN15 | Partial product accumulation | 4 |
| TN7 | Decimal fixed point | 13 | TN16 | Carry-save addition | 22 |
| TN8 | IEEE 754 | 23 | TN17 | Binary integer decimal | 5 |
| TN9 | Signed digit/SD | 12 | | | |

TN: Taxonomy Node

Table 6 provides the data for $n_x$ and $n_y$. For $n_{x,y}$, we create Table 7 populating it with the number of articles containing both $T_i$ and $T_j$; $i, j \in [1, 17]$. Therefore, Table 7 provides the numerator of Eq. 3. We then generate the denominator of Eq. 3 and populate Table 8. Table 8 basically contains the values of $\sqrt{n_x}\sqrt{n_y}$. Using Tables 7 and 8, Table 9 is populated which provides the cosine similarity index for two nodal terms, i.e., the measure of similarity between two keywords/phrases.

## 3.2 Taxonomy of Decimal Multiplier Architectures

The taxonomy is generated using Table 9. Let us assume "$i$" to be the row index and "$j$" to be the column index. Each TNi is scanned horizontally row-wise and the TNj giving the highest value is assigned the child of TNi. Since values in Table 9 are mirror images along the diagonal, hence, once a certain combination of TNi TNj has already been considered, it is not considered when $i = j$ and $j = i$. In such cases, the TNi becomes the final leaf node. For example, TN6 has the highest similarity with TN8, i.e., 0.72. Hence, TN6 is considered to be the parent of TN8. Now, while considering TN8, it fetches the highest similarity value with TN6 only; hence, this relation is discarded and TN8 is considered to be a final leaf node. All the rows are scanned $\forall i \in [1, 17]$ and the taxonomy is as follows:

**Table 7** Term—*Co-occurrence* matrix: Weightage for two nodal terms in an article

| | TN1 | TN2 | TN3 | TN4 | TN5 | TN6 | TN7 | TN8 | TN9 | TN10 | TN11 | TN12 | TN13 | TN14 | TN15 | TN16 | TN17 |
|---|---|---|---|---|---|---|---|---|---|---|---|---|---|---|---|---|---|
| TN1 | – | 10 | 5 | 32 | 1 | 17 | 11 | 12 | 12 | 9 | 23 | 5 | 9 | 8 | 3 | 19 | 1 |
| TN2 | 10 | – | 3 | 8 | 0 | 2 | 1 | 2 | 3 | 1 | 4 | 1 | 2 | 2 | 1 | 3 | 0 |
| TN3 | 5 | 3 | – | 3 | 0 | 1 | 1 | 1 | 0 | 1 | 3 | 1 | 1 | 1 | 1 | 2 | 0 |
| TN4 | 32 | 8 | 3 | – | 0 | 10 | 6 | 8 | 8 | 6 | 14 | 4 | 7 | 6 | 3 | 12 | 0 |
| TN5 | 1 | 0 | 0 | 0 | – | 3 | 1 | 1 | 0 | 1 | 1 | 2 | 0 | 0 | 0 | 0 | 0 |
| TN6 | 17 | 2 | 1 | 10 | 3 | – | 10 | 19 | 4 | 8 | 7 | 17 | 3 | 0 | 2 | 11 | 5 |
| TN7 | 11 | 1 | 1 | 6 | 1 | 10 | – | 8 | 0 | 2 | 3 | 7 | 3 | 0 | 3 | 9 | 0 |
| TN8 | 12 | 2 | 1 | 8 | 1 | 19 | 8 | – | 1 | 7 | 9 | 14 | 2 | 1 | 2 | 7 | 4 |
| TN9 | 12 | 3 | 0 | 8 | 0 | 4 | 0 | 1 | – | 2 | 4 | 0 | 2 | 5 | 0 | 5 | 0 |
| TN10 | 9 | 1 | 1 | 6 | 1 | 8 | 2 | 7 | 2 | – | 9 | 4 | 2 | 1 | 0 | 2 | 1 |
| TN11 | 23 | 4 | 3 | 14 | 1 | 7 | 3 | 9 | 4 | 9 | – | 3 | 5 | 4 | 1 | 3 | 1 |
| TN12 | 5 | 1 | 1 | 4 | 2 | 17 | 7 | 14 | 0 | 4 | 3 | – | 2 | 0 | 2 | 6 | 3 |
| TN13 | 9 | 2 | 1 | 7 | 0 | 3 | 3 | 2 | 2 | 2 | 5 | 2 | – | 4 | 2 | 4 | 5 |
| TN14 | 8 | 2 | 1 | 6 | 0 | 0 | 0 | 1 | 5 | 1 | 4 | 0 | 4 | – | 0 | 3 | 0 |
| TN15 | 3 | 1 | 1 | 3 | 0 | 2 | 3 | 2 | 0 | 0 | 1 | 2 | 2 | 0 | – | 3 | 0 |
| TN16 | 19 | 3 | 2 | 12 | 0 | 11 | 9 | 7 | 5 | 2 | 3 | 6 | 4 | 3 | 3 | – | 1 |
| TN17 | 1 | 0 | 0 | 0 | 0 | 5 | 0 | 4 | 0 | 1 | 1 | 3 | 5 | 0 | 0 | 1 | – |

**Table 8** Multiplicative data ($\sqrt{n_x}\sqrt{n_y}$)

| | TN1 | TN2 | TN3 | TN4 | TN5 | TN6 | TN7 | TN8 | TN9 | TN 10 | TN 11 | TN 12 | TN 13 | TN 14 | TN 15 | TN 16 | TN 17 |
|---|---|---|---|---|---|---|---|---|---|---|---|---|---|---|---|---|---|
| TN1 | – | 24.29 | 17.18 | 45.44 | 13.30 | 42.07 | 27.69 | 36.84 | 26.61 | 27.69 | 41.36 | 33.48 | 25.48 | 23.04 | 15.36 | 36.03 | 17.18 |
| TN2 | 24.29 | – | 7.07 | 18.71 | 5.48 | 17.32 | 11.40 | 15.17 | 10.95 | 11.40 | 17.03 | 13.78 | 10.49 | 9.49 | 6.32 | 14.83 | 7.07 |
| TN3 | 17.18 | 7.07 | – | 13.23 | 3.87 | 12.25 | 8.06 | 10.72 | 7.75 | 8.06 | 12.04 | 9.75 | 7.42 | 6.71 | 4.47 | 10.49 | 5.00 |
| TN4 | 45.44 | 18.71 | 13.23 | – | 10.25 | 32.40 | 21.33 | 28.37 | 20.49 | 21.33 | 31.86 | 25.79 | 19.62 | 17.75 | 11.83 | 27.75 | 13.23 |
| TN5 | 13.30 | 5.48 | 3.87 | 10.25 | – | 9.49 | 6.24 | 8.31 | 6.00 | 6.24 | 9.33 | 7.55 | 5.74 | 5.20 | 3.46 | 8.12 | 3.87 |
| TN6 | 42.07 | 17.32 | 12.25 | 32.40 | 9.49 | – | 19.75 | 26.27 | 18.97 | 19.75 | 29.50 | 23.87 | 18.17 | 16.43 | 10.95 | 25.69 | 12.25 |
| TN7 | 27.69 | 11.40 | 8.06 | 21.33 | 6.24 | 19.75 | – | 17.29 | 12.49 | 13.00 | 19.42 | 15.72 | 11.96 | 10.82 | 7.21 | 16.91 | 8.06 |
| TN8 | 36.84 | 15.17 | 10.72 | 28.37 | 8.31 | 26.27 | 17.29 | – | 16.61 | 17.29 | 25.83 | 20.90 | 15.91 | 14.39 | 9.59 | 22.49 | 10.72 |
| TN9 | 26.61 | 10.95 | 7.75 | 20.49 | 6.00 | 18.97 | 12.49 | 16.61 | – | 12.49 | 18.65 | 15.10 | 11.49 | 10.39 | 6.93 | 16.25 | 7.75 |
| TN10 | 27.69 | 11.40 | 8.06 | 21.33 | 6.24 | 19.75 | 13.00 | 17.29 | 12.49 | – | 19.42 | 15.72 | 11.96 | 10.82 | 7.21 | 16.91 | 8.06 |
| TN11 | 41.36 | 17.03 | 12.04 | 31.86 | 9.33 | 29.50 | 19.42 | 25.83 | 18.65 | 19.42 | – | 23.47 | 17.86 | 16.16 | 10.77 | 25.26 | 12.04 |
| TN12 | 33.48 | 13.78 | 9.75 | 25.79 | 7.55 | 23.87 | 15.72 | 20.90 | 15.10 | 15.72 | 23.47 | – | 14.46 | 13.08 | 8.72 | 20.45 | 9.75 |
| TN13 | 25.48 | 10.49 | 7.42 | 19.62 | 5.74 | 18.17 | 11.96 | 15.91 | 11.49 | 11.96 | 17.86 | 14.46 | – | 9.95 | 6.63 | 15.56 | 7.42 |
| TN14 | 23.04 | 9.49 | 6.71 | 17.75 | 5.20 | 16.43 | 10.82 | 14.39 | 10.39 | 10.82 | 16.16 | 13.08 | 9.95 | – | 6.00 | 14.07 | 6.71 |
| TN15 | 15.36 | 6.32 | 4.47 | 11.83 | 3.46 | 10.95 | 7.21 | 9.59 | 6.93 | 7.21 | 10.77 | 8.72 | 6.63 | 6.00 | – | 9.38 | 4.47 |
| TN16 | 36.03 | 14.83 | 10.49 | 27.75 | 8.12 | 25.69 | 16.91 | 22.49 | 16.25 | 16.91 | 25.26 | 20.45 | 15.56 | 14.07 | 9.38 | – | 10.49 |
| TN17 | 17.18 | 7.07 | 5.00 | 13.23 | 3.87 | 12.25 | 8.06 | 10.72 | 7.75 | 8.06 | 12.04 | 9.75 | 7.42 | 6.71 | 4.47 | 10.49 | – |

**Table 9** Cosine similarity index

|  | TN1 | TN2 | TN3 | TN4 | TN5 | TN6 | TN7 | TN8 | TN9 | TN10 | TN11 | TN12 | TN13 | TN14 | TN15 | TN16 | TN17 |
|---|---|---|---|---|---|---|---|---|---|---|---|---|---|---|---|---|---|
| TN1 | – | 0.41 | 0.29 | 0.70 | 0.08 | 0.40 | 0.40 | 0.33 | 0.45 | 0.32 | 0.56 | 0.15 | 0.35 | 0.35 | 0.20 | 0.53 | 0.06 |
| TN2 | 0.41 | – | 0.42 | 0.43 | 0.00 | 0.12 | 0.09 | 0.13 | 0.27 | 0.09 | 0.23 | 0.07 | 0.19 | 0.21 | 0.16 | 0.20 | 0.00 |
| TN3 | 0.29 | 0.42 | – | 0.23 | 0.00 | 0.08 | 0.12 | 0.09 | 0.00 | 0.12 | 0.25 | 0.10 | 0.13 | 0.15 | 0.22 | 0.19 | 0.00 |
| TN4 | 0.70 | 0.43 | 0.23 | – | 0.00 | 0.31 | 0.28 | 0.28 | 0.39 | 0.28 | 0.44 | 0.16 | 0.36 | 0.34 | 0.25 | 0.43 | 0.00 |
| TN5 | 0.08 | 0.00 | 0.00 | 0.00 | – | 0.32 | 0.16 | 0.12 | 0.00 | 0.16 | 0.11 | 0.26 | 0.00 | 0.00 | 0.00 | 0.00 | 0.00 |
| TN6 | 0.40 | 0.12 | 0.08 | 0.31 | 0.32 | – | 0.51 | 0.72 | 0.21 | 0.41 | 0.24 | 0.71 | 0.17 | 0.00 | 0.18 | 0.43 | 0.41 |
| TN7 | 0.40 | 0.09 | 0.12 | 0.28 | 0.16 | 0.51 | – | 0.46 | 0.00 | 0.15 | 0.15 | 0.45 | 0.25 | 0.00 | 0.42 | 0.53 | 0.00 |
| TN8 | 0.33 | 0.13 | 0.09 | 0.28 | 0.12 | 0.72 | 0.46 | – | 0.06 | 0.40 | 0.35 | 0.67 | 0.13 | 0.07 | 0.21 | 0.31 | 0.37 |
| TN9 | 0.45 | 0.27 | 0.00 | 0.39 | 0.00 | 0.21 | 0.00 | 0.06 | – | 0.16 | 0.21 | 0.00 | 0.17 | 0.48 | 0.00 | 0.31 | 0.00 |
| TN10 | 0.32 | 0.09 | 0.12 | 0.28 | 0.16 | 0.41 | 0.15 | 0.40 | 0.16 | – | 0.46 | 0.25 | 0.17 | 0.09 | 0.00 | 0.12 | 0.12 |
| TN11 | 0.56 | 0.23 | 0.25 | 0.44 | 0.11 | 0.24 | 0.15 | 0.35 | 0.21 | 0.46 | – | 0.13 | 0.28 | 0.25 | 0.09 | 0.12 | 0.08 |
| TN12 | 0.15 | 0.07 | 0.10 | 0.16 | 0.26 | 0.71 | 0.45 | 0.67 | 0.00 | 0.25 | 0.13 | – | 0.14 | 0.00 | 0.23 | 0.29 | 0.31 |
| TN13 | 0.35 | 0.19 | 0.13 | 0.36 | 0.00 | 0.17 | 0.25 | 0.13 | 0.17 | 0.17 | 0.28 | 0.14 | – | 0.40 | 0.30 | 0.26 | 0.67 |
| TN14 | 0.35 | 0.21 | 0.15 | 0.34 | 0.00 | 0.00 | 0.00 | 0.07 | 0.48 | 0.09 | 0.25 | 0.00 | 0.40 | – | 0.00 | 0.21 | 0.00 |
| TN15 | 0.20 | 0.16 | 0.22 | 0.25 | 0.00 | 0.18 | 0.42 | 0.21 | 0.00 | 0.00 | 0.09 | 0.23 | 0.30 | 0.00 | – | 0.32 | 0.00 |
| TN16 | 0.53 | 0.20 | 0.19 | 0.43 | 0.00 | 0.43 | 0.53 | 0.31 | 0.31 | 0.12 | 0.12 | 0.29 | 0.26 | 0.21 | 0.32 | – | 0.10 |
| TN17 | 0.06 | 0.00 | 0.00 | 0.00 | 0.00 | 0.41 | 0.00 | 0.37 | 0.00 | 0.12 | 0.08 | 0.31 | 0.67 | 0.00 | 0.00 | 0.10 | – |

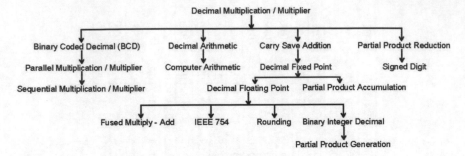

**Fig. 5** Taxonomy tree for decimal multiplier research

**T1.** Decimal Multiplication/Multiplier (Term TN1)

    **T1.1** Binary Coded Decimal (BCD) (Term TN4)
        **T1.1.1** Parallel Multiplication/Multiplier (Term TN2)
        **T1.1.2** Sequential Multiplication/Multiplier (Term TN3)
    **T1.2** Decimal Arithmetic (Term TN11)
        **T1.2.1** Computer Arithmetic (Term TN10)
    **T1.3** Carry-Save Addition (Term TN16)
        **T1.3.1** Decimal Fixed Point (Term TN7)
          **T1.3.1.1** Decimal Floating Point (Term TN6)
    **T1.3.1.1.1** Fused Multiply–Add (Term TN5)
    **T1.3.1.1.2** IEEE 754 (Term TN8)
    **T1.3.1.1.3** Rounding (Term TN12)
    **T1.3.1.1.4** Binary Integer Decimal (Term TN17)
        **T1.3.1.1.4.1** Partial Product Generation (Term TN13)
    **T1.3.1.2** Partial Product Accumulation (Term TN15)

    **T1.4** Partial Product Reduction (Term TN14)

    **T1.4.1** Signed Digit (Term TN9)

  The above taxonomy is visualized in a tree form in Fig. 5.

## 3.3  Mapping of Taxonomy and Published Literature

The taxonomy generated in the previous section is mapped to the articles. Table 10 provides the document Ids of all the articles that contain a certain Keyword/Term.

**Table 10** Taxonomy nodal terms versus document Ids (Published Literature)

| Term Id | Document Id | Term Id | Document Id |
|---------|-------------|---------|-------------|
| TN1 | PL1, PL2, PL3, PL4, PL5, PL7, PL8, PL9, PL10, PL13, PL14, PL15, PL16, PL17, PL18, PL19, PL20, PL21, PL22, PL23, PL24, PL25, PL27, PL28, PL29, PL32, PL33, PL37, PL38, PL39, PL40, PL41, PL42, PL43, PL44, PL46, PL47, PL49, PL51, PL52, PL54, PL56, PL57, PL58, PL60, PL61, PL62, PL63, PL64, PL65, PL66, PL67 | TN10 | PL3, PL21, PL22, PL26, PL29, PL30, PL44, PL51, PL53, PL61, PL62 |
| TN2 | PL1, PL2, PL8, PL9, PL10, PL18, PL24, PL39, PL40, PL44 | TN11 | PL1, PL9, PL10, PL11, PL12, PL20, PL21, PL22, PL26, PL27, PL28, PL29, PL33, PL35, PL41, PL42, PL43, PL44, PL47, PL50, PL52, PL58, PL61, PL62, PL63 |
| TN3 | PL1, PL2, PL8, PL22, PL27 | TN12 | PL2, PL5, PL6, PL12, PL13, PL15, PL26, PL30, PL31, PL34, PL36, PL45, PL48, PL53, PL57, PL59, PL64 |
| TN4 | PL1, PL2, PL3, PL5, PL8, PL16, PL17, PL18, PL19, PL21, PL24, PL25, PL28, PL32, PL33, PL35, PL37, PL39, PL40, PL41, PL42, PL44, PL46, PL47, PL48, PL49, PL50, PL51, PL52, PL54, PL57, PL61, PL62 | TN13 | PL5, PL9, PL10, PL22, PL25, PL29, PL35, PL46, PL47, PL48 |
| TN5 | PL45, PL59 | TN14 | PL9, PL16, PL22, PL35, PL39, PL41, PL49 |
| TN6 | PL2, PL3, PL5, PL7, PL12, PL13, PL15, PL23, PL25, PL26, PL30, PL31, PL36, PL40, PL45, PL48, PL51, PL52, PL53, PL55, PL56, PL57, PL59, PL62, PL63, PL64 | TN15 | PL2, PL13, PL46, PL47 |
| TN7 | PL2, PL5, PL7, PL13, PL15, PL17, PL19, PL23, PL26, PL47, PL48, PL64 | TN16 | PL2, PL3, PL5, PL6, PL7, PL13, PL14, PL17, PL19, PL23, PL27, PL31, PL32, PL39, PL40, PL47, PL48, PL25, PL56, PL58 |
| TN8 | PL2, PL5, PL12, PL13, PL23, PL26, PL30, PL31, PL36, PL42, PL44, PL48, PL52, PL53, PL57, PL59, PL62, PL63, PL64 | TN17 | PL31, PL36, PL53, PL55, PL63 |
| TN9 | PL3, PL9, PL16, PL20, PL39, PL40, PL49, PL51, PL52, PL65 | | |

Basically, Table 10 provides all the articles related to a certain node of interest in the taxonomy. This table is of particular research interest as it provides all the relevant publications pertaining to a specific subarea of decimal multiplier research. For example, TN12 provides all the published articles related to rounding architecture proposals for decimal multipliers in literature. Table 11 provides the inverse mapping of Table 10. Table 11 is populated using keywords mapped to document Ids. Table 11 provides percentage coverage of the taxonomy by a given article. Therefore, PL2 covers 59% of the nodes in the taxonomy and going by Table 1, research content of [4] justifies the findings.

**Table 11** Reverse mapping: Document Ids versus nodal terms

| DiD | Term Id | % | DiD | Term Id | % | DiD | Term Id | % |
|---|---|---|---|---|---|---|---|---|
| PL1 | TN1, TN2, TN3, TN4, TN11 | 29.5 | PL24 | TN1, TN2, TN4 | 17.7 | PL47 | TN1, TN4, TN17, TN11, TN13, TN15, TN16 | 41.3 |
| PL2 | TN1, TN2, TN3, TN4, TN6, TN7, TN8, TN12, TN15, TN16 | 59 | PL25 | TN1, TN4, TN6, TN13 | 23.6 | PL48 | TN4, TN6, TN7, TN8, TN12, TN13, TN16 | 41.3 |
| PL3 | TN1, TN4, TN6, TN9, TN10, TN16 | 35.4 | PL26 | TN6, TN7, TN8, TN10, TN11, TN12 | 29.5 | PL49 | TN1, TN4, TN9, TN14 | 23.6 |
| PL4 | TN1 | 5.9 | PL27 | TN1, TN4, TN11, TN16 | 23.6 | PL50 | TN4, TN11 | 11.8 |
| PL5 | TN1, TN4, TN6, TN7, TN8, TN12, TN13, TN16 | 47.2 | PL28 | TN1, TN4, TN11 | 17.7 | PL51 | TN1, TN4, TN6, TN9, TN10, TN16 | 35.4 |
| PL6 | TN12, TN16 | 11.8 | PL29 | TN1, TN10, TN11, TN13 | 23.6 | PL52 | TN1, TN4, TN6, TN8, TN9, TN11 | 35.4 |
| PL7 | TN1, TN6, TN7, TN16 | 23.6 | PL30 | TN6, TN8, TN10, TN12 | 23.6 | PL53 | TN6, TN8, TN10, TN12, TN17 | 29.5 |
| PL8 | TN1, TN2, TN3, TN4 | 23.6 | PL31 | TN6, TN8, TN12, TN16, TN17 | 29.5 | PL54 | TN1, TN4 | 11.8 |
| PL9 | TN1, TN2, TN9, TN11, TN13, TN14 | 35.4 | PL32 | TN1, TN4 | 11.8 | PL55 | TN6, TN17 | 11.8 |
| PL10 | TN1, TN2, TN11, TN13 | 23.6 | PL33 | TN1, TN4, TN11 | 17.7 | PL56 | TN1, TN6, TN16 | 17.7 |

(continued)

**Table 11** (continued)

| DiD | Term Id | % | DiD | Term Id | % | DiD | Term Id | % |
|---|---|---|---|---|---|---|---|---|
| PL11 | TN11 | 5.9 | PL34 | TN12 | 5.9 | PL57 | TN1, TN4, TN6, TN8, TN12 | 29.5 |
| PL12 | TN6, TN8, TN11, TN12 | 23.6 | PL35 | TN4, TN11, TN13, T14 | 23.6 | PL58 | TN1, TN11, TN16 | 17.7 |
| PL13 | TN1, TN6, TN7, TN8, TN12, TN15, TN16 | 41.3 | PL36 | TN6, TN8, TN12, TN17 | 23.6 | PL59 | TN6, TN8, TN12 | 17.7 |
| PL14 | TN1, TN16 | 11.8 | PL37 | TN1, TN4 | 11.8 | PL60 | TN1 | 5.9 |
| PL15 | TN1, TN6, TN7, TN12 | 23.6 | PL38 | TN1 | 5.9 | PL61 | TN1, TN4, TN10, TN11 | 23.6 |
| PL16 | TN1, TN4, TN9, TN14 | 23.6 | PL39 | TN1, TN2, TN4, TN9, TN14, TN16 | 35.4 | PL62 | TN1, TN4, TN6, TN8, TN10, TN11 | 35.4 |
| PL17 | TN1, TN4, TN7, TN16 | 23.6 | PL40 | TN1, TN2, TN4, TN6, TN9, TN16 | 35.4 | PL63 | TN1, TN6, TN8, TN11 | 23.6 |
| PL18 | TN1, TN2, TN4 | 17.7 | PL41 | TN1, TN4, TN11, TN14 | 23.6 | PL64 | TN1, TN6, TN7, TN8, TN12 | 29.5 |
| PL19 | TN1, TN4, TN7, TN16 | 23.6 | PL42 | TN1, TN4, TN8, TN11 | 23.6 | PL65 | TN1, TN9 | 11.8 |
| PL20 | TN1, TN9, TN11 | 17.7 | PL43 | TN1, TN11 | 11.8 | PL66 | TN1 | 5.9 |
| PL21 | TN1. TN4, TN10, TN11 | 23.6 | PL44 | TN1, TN2, TN4, TN8, TN10, TN11 | 35.4 | PL67 | TN1 | 5.9 |
| PL22 | TN1, TN3, TN10, TN11, TN13, TN14 | 35.4 | PL45 | TN5, TN6, TN12 | 17.7 | | | |
| PL23 | TN1, TN6, TN7, TN8, TN16 | 29.5 | PL46 | TN1, TN4, TN13, TN15 | 23.6 | | | |

# 4 Conclusion

We have provided taxonomy for decimal multiplier research in this study. The prime objective of the taxonomy is to present a platform where node-wise research can be conducted in future, i.e., beginning research in decimal rounding (T1.3.1.1.3, Term TN12) will require an initial study of published literature provided in Table 10 alongside TN12. This study classifies articles in decimal multiplier research into different nodes of taxonomy so as to minimize the literature survey duration for initiating research in a particular node within the domain. Table 11 provides the % content of taxonomy in a certain published literature. Hence, articles having taxonomy coverage of more than 25% can be termed as state-of-the-art articles and referred to as foundation literature for research on decimal multiplier architectures.

The present study can be further explored using more robust algorithms, i.e., Google similarity distance, etc. It can also be extended to form a sub-domain within taxonomy for decimal arithmetic architectures.

# References

1. Schmid, H.: Decimal Computation. Wiley (1974)
2. Richards, R.: Arithmetic Operations in Digital Computers. D. Van Nostrand Company, Inc. (1955)
3. Vazquez, A., Antelo, E., Montuschi, P.: Improved design of high-performance parallel decimal multipliers. IEEE Trans. Comput. 59(5), 679–693 (2010)
4. Erle, M., Hickmann, B., Schulte, M.: Decimal floating-point multiplication. IEEE Trans. Comput. 58(7), 902–916 (2009)
5. Cornea, M., Anderson, C., Harrison, J., Tang, P., Schneider, E., Tsen, C.: A software implementation of the IEEE 754R decimal floating-point arithmetic using the binary encoding format. In: 18th IEEE Symposium on Computer Arithmetic, 2007. ARITH '07, Montepellier, pp. 29–37 (2007)
6. Cornea, M.: Intel® Decimal Floating-Point Math Library. Intel® Corporation (2011)
7. JAVA, Sun Microsystems: Class BigDecimal Documentation. JAVA (1996)
8. Cowlishaw, M.: The decNumber C Library, Version 3.68., IBM (January 2010)
9. Simington, R.: The intel 8087 numerics processor extension. BYTE Mag. 8(4), 154 (1983)
10. Cowlishaw, M.: Decimal floating-point: algorism for computers. In: Proceedings of the 16th IEEE Symposium on Computer Arithmetic (ARITH'03), pp. 104–111 (2003)
11. Erle, M., Schulte, M., Linebarger, J.: Potential speedup using decimal floating-point hardware. In: Conference Record of the Thirty-Sixth Asilomar Conference on Signals, Systems and Computers, 2002, Pacific Grove, CA, USA, vol. 2, pp. 1073–1077, Nov 2002
12. Webb, C.: IBM z10: the next-generation mainframe microprocessor. IEEE Micro 28(2), 19–29 (2008)
13. Raafat, R., Abdel-Majeed, A., Samy, R., ElDeeb, T., Farouk, Y., Elkhouly, M., Fahmy, H.: A decimal fully parallel and pipelined floating point multiplier. In: 42nd Asilomar Conference on Signals, Systems and Computers, Pacific Grove, pp. 1800–1804 (2008)
14. Fahmy, H., ElDeeb, T., Hassan, M.: Decimal floating point for future processors. In: International Conference on Microelectronics (ICM), Cairo, pp. 443–446 (2010)
15. Cowlishaw, M.: Densely packed decimal encoding. Comput. Digit. Techn. 149(3), 102–104 (2002)

16. IEEE Computer Society: IEEE Standards for Floating-Point Arithmetic 754-2008, IEEE. Aug 2008
17. Quach, N., Takagi, N., Flynn, M.: Systematic IEEE rounding method for high-speed floating-point multipliers. IEEE Trans. Very Larg. Scale Integr. VLSI Syst. **12**(5), 511–521 (2004)
18. Even, G., Seidel, P.: A comparison of three rounding algorithms for IEEE floating-point multiplication. IEEE Trans. Comput. **49**(7), 638–650 (2000)
19. Wang, L.-K., Schulte, M.: Decimal floating-point adder and multifunction unit with injection-based rounding. In: 18th IEEE Symposium on Computer Arithmetic, 2007. ARITH '07, Montepellier, pp. 56–68 (2007)
20. Wang, L.-K., Schulte, M., Thompson, J., Jairam, N.: Hardware designs for decimal floating-point addition and related operations. IEEE Trans. Comput. **58**(3), 322–335 (2009)
21. Tsen, C., Gonzalez-Navarro, S., Schulte, M., Hickmann, B., Compton, K.: A combined decimal and binary floating-point multiplier. In: 2009 20th IEEE International Conference on Application-specific Systems, Architectures and Processors, Boston, MA, pp. 8–15 (2009)
22. Tsen, C., Schulte, M., Gonzalez-Navarro, S.: Hardware design of a binary integer decimal-based IEEE P754 rounding unit. In: IEEE International Conference on Application-specific Systems, Architectures and Processors (ASAP) 2007, Montreal, Que, pp. 115–121 (2007)
23. Camina, S.: A comparison of taxonomy generation techniques using Bibliometric. EECS Thesis, Massachusetts Institute of technology (2010)
24. Guardia, C.: Implementation of a fully pipelined BCD multiplier in FPGA. In: VIII Southern Conference on Programmable Logic (SPL), 2012, Bento Goncalves, pp. 1–6 (2012)
25. Carlough, S., Schwarz, E.: Decimal Multiplication using Digit Recoding. US, Patent US 7136893 B2, US. 14 Nov 2006
26. Baesler, M., Teufel, T.: FPGA implementation of a decimal floating-point accurate scalar product unit with a parallel fixed-point multiplier. In: International Conference on Reconfigurable Computing and FPGAs 2009, Quintana Roo, pp. 6–11 (2009)
27. Erle, M., Schulte, M.: Decimal multiplication via carry-save addition. In: Proceedings IEEE International Conference on Application-Specific Systems, Architectures, and Processors, pp. 348–358 (2003)
28. Croy, J.: Improved arrangement of a decimal multiplier. IRE Trans. Electron. Comput. EC-**9**(2), 263 (1960)
29. Han, L., Ko, S.-B.: High-speed parallel decimal multiplication with redundant internal encodings. IEEE Trans. Comput. **62**(5), 956–968 (2013)
30. Castellanos, I., Stine, J.: Decimal partial product generation architectures. In: 51st Midwest Symposium on Circuits and Systems 2008, Knoxville, TN, pp. 962–965 (2008)
31. Bozdas, K., Alkar, A.: Analysis on the column sum boundaries of decimal array multipliers. In: IEEE 55th International Midwest Symposium on Circuits and Systems (MWSCAS) 2012, Boise, ID, pp. 318–321 (2012)
32. Erle, M., Schulte, M., Hickmann, B.: Decimal floating-point multiplication via carry-save addition. In: 18th IEEE Symposium on Computer Arithmetic, 2007. ARITH '07, Montepellier, pp. 46–55. June 2007
33. Dadda, L., Nannarelli, A.: A variant of a Radix-10 combinational multiplier. In: IEEE International Symposium on Circuits and Systems (ISCAS 2008), pp. 3370–3373 (2008)
34. Hickman, B., Krioukov, A., Schulte, M., Erle, M.: A parallel IEEE P754 decimal floating-point multiplier. In: 25th International Conference on Computer Design, 2007. ICCD 2007. Lake Tahoe, CA, pp. 296–303 (2007)
35. Gorgin, S., Jaberipur, G.: Sign-magnitude encoding for efficient VLSI realization of decimal multiplication. IEEE Trans. Very Larg. Scale Integr. (VLSI) Syst. (99), 1–13 (2016)
36. Erle, M., Hickmann, B.: Combined binary/decimal fixed-point multiplier and method. US, Patent US8577952 B2, US. 5 Nov 2013
37. Dadda, L., Pisoni, M., Santambrogio, M.: A parallel-serial decimal multiplier architecture. In: IEEE 15th International Conference on Computational Science and Engineering (CSE), 2012, Nicosia, pp. 310–317 (2012)

38. Hickmann, B., Schulte, M., Erle, M.: Improved combined binary/decimal fixed-point multipliers. In: IEEE International Conference on Computer Design, 2008. ICCD 2008. Lake Tahoe, CA, pp. 87–94 (2008)
39. Gorgin, S., Jaberipur, G.: Fully redundant decimal arithmetic. In: 2009 19th IEEE International Symposium on Computer Arithmetic, pp. 145–152 (2009)
40. Jaberipur, G., Kaivani, A.: Binary-coded decimal digit multipliers. IET Comput. Digit. Tech. 1(4), 377–381 (2007)
41. Jaberipur, G., Kaivani, A.: Improving the speed of parallel decimal multiplication. IEEE Trans. Comput. 58(11), 1539–1552 (2009)
42. James, R., Jacob, K., Sasi, S.: High performance, low latency double digit decimal multiplier on ASIC and FPGA. In: World Congress on Nature and Biologically Inspired Computing, 2009. NaBIC 2009, Coimbatore, pp. 1445–1450 (2009)
43. James, R., Shahana, T., Jacob, K., Sasi, S.: Decimal multiplication using compact BCD multiplier. In: Electronic Design, 2008. ICED 2008. penang, pp. 1–6. Dec 2008
44. James, R., Shahana, T., Jacob, P., Sasi, S.: Fixed point decimal multiplication using RPS algorithm. In: IEEE International Symposium on Parallel and Distributed Processing with Applications 2008, Sydney, NSW, pp. 343–350 (2008)
45. Kaivani, A., Han, L., Ko, S.-B.: Improved design of high-frequency sequential decimal multipliers. Electron. Lett. Inst. Eng. Technol. 50(7), 558–560 (2014)
46. Kenney, R., Schulte, M., Erle, M.: A high-frequency decimal multiplier. In: Proceedings IEEE International Conference on Computer Design: VLSI in Computers and Processors, 2004. ICCD 2004, pp. 26–29 (2004)
47. Lin, K., Chiu, Y., Lin, T.-H.: A decimal squarer with efficient partial product generation. In: 18th IEEE/IFIP International Conference on VLSI and System-on-Chip 2010, Madrid, pp. 213–218 (2010)
48. Navarro, S., Tsen, C., Schulte, M.: Binary integer decimal-based floating-point multiplication. IEEE Trans. Comput. 62(7), 1460–1466 (2013)
49. Ohtsuki, T., Oshima, Y., Ishikawa, S., Yabe, H., Fukuta, M.: Apparatus for decimal multiplication. US, Patent US 4677583 A, US. 30 June 1987
50. Osama, D., Khaleel, A., Tulic, N., Mhaidat, K.: FPGA implementation of binary coded decimal digit adders and multipliers. In: 8th International Symposium on Mechatronics and its Applications (ISMA), 2012, Sharjah, pp. 1–5 (2012)
51. Sutter, G., Todorovich, E., Bioul, G., Vazquez, M., Deschamps, J.: FPGA implementations of BCD multipliers. In: International Conference on Reconfigurable Computing and FPGAs, 2009. ReConFig '09. Quintana Roo, pp. 36–41 (2009)
52. Ueda, T.: Decimal multiplying assembly and multiply module. US, Patent US 5379245, US. Jan 1995
53. Vázquez, Á., Antelo, E., Bruguera, J.: Fast Radix-10 multiplication using redundant BCD codes. IEEE Trans. Comput. 63(8), 1902–1914 (2014)
54. Veeramachaneni, S., Srinivas, M.: Novel high-speed architecture for 32-Bit binary coded decimal (BCD) multiplier. In: 2008 International Symposium on Communications and Information Technologies, 2008. ISCIT, Lao, pp. 543–546 (2008)
55. Véstias, M., Neto, H.: Iterative decimal multiplication using binary arithmetic. In: 2011 VII Southern Conference on VII Southern Conference on Programmable Logic (SPL). Cordoba, pp. 257–262 (2011)
56. Véstias, M., Neto, H.: Parallel decimal multipliers and squarers using Karatsuba-Ofman's algorithm. In: 15th Euromicro Conference on Digital System Design (DSD), 2012, Izmir, pp. 782–788 (2012)
57. Véstias, M., Neto, H.: Parallel decimal multipliers using binary multipliers. In: VI Southern Programmable Logic Conference (SPL), 2010, Ipojuca, pp. 73–78 (2010)
58. Wahba, A., Fahmy, H.: Area efficient and fast combined binary/decimal floating point fused multiply add unit. IEEE Trans. Comput. (99), 1 (2016)
59. Zhu, M., Baker, A., Jiang, Y.: On a parallel decimal multiplier based on hybrid 8421–5421 BCD recoding. In: 2013 IEEE 56th International Midwest Symposium on Circuits and Systems (MWSCAS), Columbus, OH, pp. 1391–1394 (2013)

60. Zhu, M., Jiang, Y.: An area-time efficient architecture for $16 \times 16$ decimal multiplications. In: Tenth International Conference on Information Technology: New Generations (ITNG), 2013, Las Vegas, NV, pp. 210–216 (2013)
61. Baesler, M., Voigt, S.-O., Teufel, T.: A decimal floating-point accurate scalar product unit with a parallel fixed-point multiplier on a virtex-5 FPGA. Int. J. Reconfigurable Comput. **2010**, 13 (2010). (Article ID 357839)
62. Cui, X., Liu, W., Wenwen, D., Lombardi, F.: A parallel decimal multiplier using hybrid binary coded decimal (BCD) codes. In: IEEE 23nd Symposium on Computer Arithmetic (ARITH) **2016** (2016)
63. Varma, C., Ahmed, S., Srinivas, M.: A decimal/binary multi-operand adder using a fast binary to decimal converter. In: 27th International Conference on VLSI Design and 2014 13th International Conference on Embedded Systems (2014)
64. Eduardo, C., Guardia, M.: Implementation of a fully pipelined BCD multiplier in FPGA. In: VIII Southern Conference on Programmable Logic (SPL) (2012)
65. Ding, H., Shu, P., Wang, X., Yang, J.: A design and implementation of decimal floating-point multiplication unit based on SOPC. In: Third International Conference on Digital Manufacturing and Automation (ICDMA) (2012)
66. Tsen, S., Gonzalez-Navarro, S., Schulte, M., Compton, K.: Hardware designs for binary integer decimal-based rounding. IEEE Trans. Comput. **60**(5), 614–627 (2011)
67. Vázquez, Á., Dinechin, F.: Efficient implementation of parallel BCD multiplication in LUT-6 FPGAs. In: International Conference on Field-Programmable Technology (FPT) (2010)
68. Navarro, S., Tsen, C., Schulte, M.: A binary integer decimal-based multiplier for decimal floating-point arithmetic. In: Forty-First Asilomar Conference on Signals, Systems and Computers, 2007. ACSSC 2007 (2007)
69. Rekha, K., Jacob, K., Sasi, S.: Performance analysis of double digit decimal multiplier on various FPGA logic families. In: 5th Southern Conference on Programmable Logic, 2009. SPL. Sao Carlos, pp. 165–170 (2009)
70. Minchola, C., Sutter, G.: A FPGA IEEE-754-2008 Decimal64 floating-point multiplier. In: International Conference on Reconfigurable Computing and FPGAs, 2009. ReConFig '09. Quintana Roo, pp. 59–64 (2009)
71. Kaivani, A., Chen, L., Ko, S.: High-frequency sequential decimal multipliers. In: 2012 IEEE International Symposium on Circuits and Systems (ISCAS). Seoul, pp. 3045–3048 (2012)
72. Lang, T., Nannarelli, A.: A Radix-10 combinational multiplier. In: Fortieth Asilomar Conference on Signals, Systems and Computers, 2006. ACSSC '06. Pacific Grove, CA, pp. 313–317 (2006)
73. Gorgin, S., Jaberipur, G., Parhami, B.: Design and evaluation of decimal array multipliers. In: 2009 Conference Record of the Forty-Third Asilomar Conference on Signals, Systems and Computers, pp. 1782–1786. IEEE. Nov 2009
74. Neto, H., Vestias, M.: Decimal multiplier on FPGA using embedded binary multipliers. In: 2008 International Conference on Field Programmable Logic and Applications, Heidelberg, pp. 197–202 (2008)
75. Gonzalez-Navarro, S., Tsen, C., Schulte, M.: A binary integer decimal-based multiplier for decimal floating-point arithmetic. In: 2007 Conference Record of the Forty-First Asilomar Conference on Signals, Systems and Computers, Pacific Grove, CA, pp. 353–357 (2007)
76. Baesler, M., Voigt, S.-O., Teufel, T.: An IEEE 754-2008 decimal parallel and pipelined FPGA floating-point multiplier. In: 2010 International Conference on Field Programmable Logic and Applications (FPL). Milano, pp. 489–495 (2010)
77. Erle, M., Schwarz, E., Schulte, M.: Decimal multiplication with efficient partial product generation. In: 17th IEEE Symposium on Computer Arithmetic, 2005. ARITH-17 2005, pp. 21–28 (2005)
78. Jouppi, N.: Wallace-tree multipliers using half and full adders. US, Patent US 6065033 A, US. 16 May 2000
79. Lehman, M.: Short-cut multiplication and division in automatic binary digital computers, with special reference to a new multiplication process. Proc. IEEE—Part B: Radio Electron. Eng. **105**(23), 496–504 (2010)

# A Synoptic Study on Fault Testing in Reversible and Quantum Circuits

Ramyadeep Dey, Paramita Bandyopadhyay, Somenath Chandra
and Ritajit Majumdar

**Abstract** Reversible computation, whose special class is quantum computation, arises from the desire to reduce power dissipation, which can be zero under ideal physical circumstances. Nowadays, error correction and fault testing are of utmost importance for the physical implementation of reversible and quantum circuits in a noisy environment. In this paper, we review various fault models in reversible and quantum circuits. In classical reversible circuits, we review (i) test pattern generation for Single Missing Gate Fault (SMGF), Partial Missing Gate Fault (PMGF), and Multiple Missing Gate Fault (MMGF) models and (ii) show that Universal Test Set (UTS) can be used to detect any of these faults. However, classical fault models do not capture all the logical failures found in quantum circuits. In quantum circuits, we review (i) depolarizing faults, initialization inaccuracy and measurement inaccuracy, and (ii) give some remedial strategies to deal with these fault models. Finally, we show that for a special class of quantum operators, detection of SMGF is sufficient to detect fault due to multiple occurrences of the gate also. Further, we argue with an example that it may not be possible to generate test patterns to detect any arbitrary SMGF in a quantum circuit.

**Keywords** Fault testing · Reversible circuits · Quantum circuits

## 1 Introduction

Consumption of energy is an important factor in digital logic design, which needs to be reduced. Landauer [10] showed that the heat dissipation is due to the loss of information during computation. Loss of each bit of information results in a dissipation of heat equal to $kTln2$, where $k$ is the Boltzmann constant and $T$ is

R. Dey · P. Bandyopadhyay · S. Chandra
Department of Computer Science & Engineering, B. P. Poddar Institute of Management & Technology, Maulana Abul Kalam Azad University of Technology, Kolkata, India

R. Majumdar (✉)
Advanced Computing & Microelectronics Unit, Indian Statistical Institute, Kolkata, India
e-mail: majumdar.ritajit@gmail.com

© Springer Nature Singapore Pte Ltd. 2018
S. K. Das and N. Chaki (eds.), *Algorithms and Applications*, Smart Innovation,
Systems and Technologies 88, https://doi.org/10.1007/978-981-10-8102-6_2

**Fig. 1** NOT Gate

the temperature in absolute scale. Though this value is pretty low for a single bit of information loss, in modern computers, millions of instructions are computed every second, and hence the heat dissipation becomes formidable. It was Bennett [2] who showed that in order to avoid that power dissipation, the computation must be reversible. This proof by Bennett has attracted the interest of many researchers toward reversible logic synthesis [14, 21, 27–29].

Evolution of a quantum bit or qubit is governed by unitary operators, and hence quantum computation is inherently reversible. Unlike a classical bit, a qubit can be in a superposition of both $|0\rangle$ *and* $|1\rangle$. A general quantum state is mathematically denoted as $\alpha |0\rangle + \beta |1\rangle$. Quantum computing promises to reduce the computational complexity of some problems [7, 23]. However, qubits are extremely prone to errors. Hence, error correction is of utmost importance for implementation of a quantum computer. This has led to study on quantum error correction [6, 22, 24] and their efficient implementation [9, 13].

In the recent times, many researchers are focusing on the testing and fault modeling for reversible and quantum circuits. In this paper, we present a review of various faults occurring on reversible and quantum circuits, and their testing. Their testability and design by k-CNOT gates are also studied. We also argue with an example that some fault models that can be corrected in classical reversible circuits may not be correctable in quantum circuits, which may prove to be a serious hindrance in the design of fault-free quantum circuits.

The rest of the paper is organized as follows—In Sect. 2, we discuss some of the basic reversible gates. In Sects. 3 and 4, we discuss the various fault models in reversible circuits and their testing, respectively. Section 5 addresses some faults in quantum circuit and in Sect. 6, we make our arguments on the impossibility of testing SMGF in quantum circuits. We conclude in Sect. 7.

## 2  Basic Reversible Gates

**NOT Gate**: NOT gate is the simplest reversible gate (Fig. 1). It is a $(1 \times 1)$ gate, i.e., there is a single input and a single output. The function of this gate is to invert the input value and is denoted as $[\{x\} \Rightarrow \{x'\}]$.

**CNOT Gate**: CNOT or Controlled-Not Gate [5] is a $(2 \times 2)$ gate (Fig. 2). One of the inputs is the control and the another one is the target. After the gate operation, the control input passes unchanged while the value of the target input is complemented iff the value of control input is 1. The function is denoted as $[\{x, y\} \Rightarrow \{x, x \oplus y\}]$.

**Toffoli Gate**: The Toffoli gate [28] is a $(3 \times 3)$ gate where there are two control lines and one target line (Fig. 3). If both the control lines are set to logical 1, then

**Fig. 2** CNOT Gate

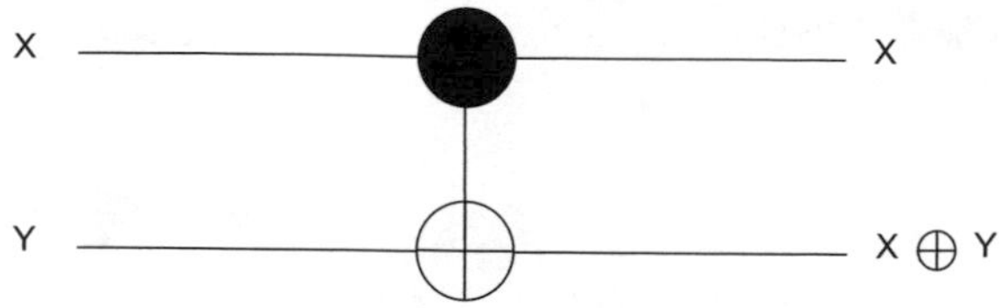

**Fig. 3** Toffoli Gate

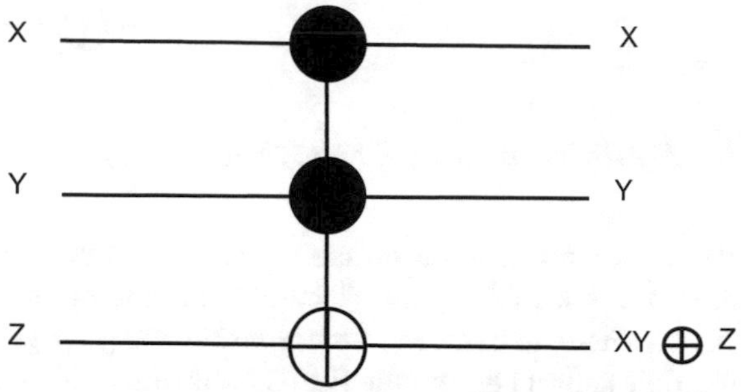

**Fig. 4** Fredkin Gate

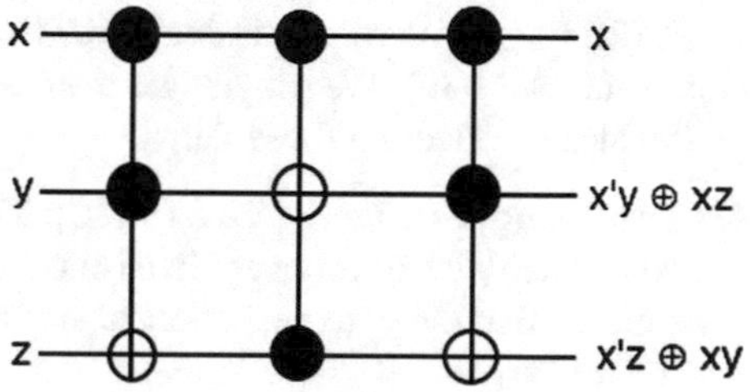

the output value is inverted, otherwise it remains same. The function is denoted as $[\{x, y, z\} \Rightarrow \{x, y, xy \oplus z\}]$. It is immediately evident that

$$NAND(x, y) = TOFFOLI(x, y, 1)$$

Since NAND gate is universal and can be implemented using TOFFOLI gate, the latter is also universal for classical reversible circuits.

**Fredkin Gate**: Fredkin gate is also a $(3 \times 3)$ gate (Fig. 4) where the second and third inputs are swapped with each other if the first input is set to 1; otherwise, it remains the same [28]. The function is denoted as $[\{x, y, z\} \Rightarrow \{x, x'y \oplus xz, x'z \oplus xy\}]$.

**Peres Gate**: Peres Gate (PG) is a $(3 \times 3)$ gate composed of two XOR gates and one AND gate (Fig. 5) [19]. The function is denoted as $[\{x, y, z \Rightarrow x, x \oplus y, xy \oplus z\}]$.

**Fig. 5** Peres Gate

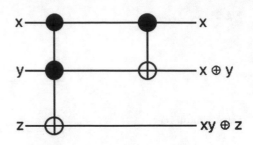

## 3   Fault Models in Reversible Circuit

We are mainly focused on the faults that occur in a reversible circuit due to the presence of k-CNOT gates ($k$ number of control lines). This is because, for such a gate, a single fault in one of the control lines propagate to the target line, leading to multiple faults in the circuit. Faults occurring on a $k-$CNOT gate can be categorized into four basic groups—(i) Single Missing Gate Fault (SMGF), (ii) Repeated Gate Fault (RGF), (iii) Partial Missing Gate Fault (PMGF), and (iv) Multiple Missing Gate Fault (MMGF). We provide an overview of these faults individually and touch on the idea of detecting these faults.

*Single Missing Gate Fault (SMGF)* [20]: This type of fault occurs when one of the k-CNOT gates (Fig. 6) is not applied (or the gate fails, hence acting simply as a wire). This means that the gate gets short or missing. Such a fault can be detected by the test vector $\{x_1, x_2, x_3\} = \{0, 1, 1\}$.

*Repeated Gate Fault (RGF)* [20]: This fault occurs when there is one or more unwanted repetition of a k-CNOT gate (Fig. 7). It can be detected by the test vector $\{x_1, x_2, x_3\} = \{0, 1, 1\}$ which is same as that of single missing gate fault.

*Partial Missing Gate Fault (PMGF)* [20]: Such a fault occurs when gate pulses are partially misaligned or mistuned (Fig. 8). For example, one of the control signals may not work properly, thus incorporating fault in both the control line and the target line. Such a fault changes a k-CNOT gate into a p-CNOT gate, $p < k$. Hence, this fault is also called $(k - p)$-th order PMGF. It is important to note that SMGF is a special case of PMGF (0-th order PMGF). The test set $\{x_1, x_2, x_3\} = \{1, 0, 1\}$ detects this fault.

*Multiple Missing Gate Fault (MMGF)* [20]: This fault occurs (Fig. 9) when two or more consecutive k-CNOT gates are not applied. This fault is detected by the test vector $\{x_1, x_2, x_3\} = \{0, 1, 1\}$.

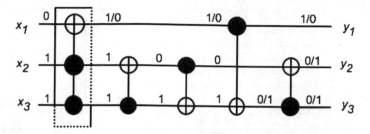

**Fig. 6** Single Missing Gate Fault

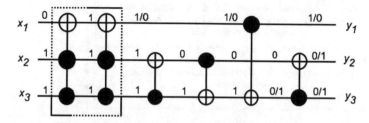

**Fig. 7** Repeated Gate Fault

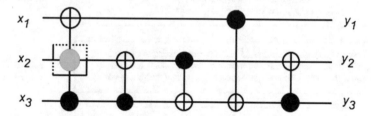

**Fig. 8** Partial Missing Gate Fault

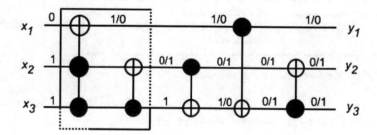

**Fig. 9** Multiple Missing Gate Fault

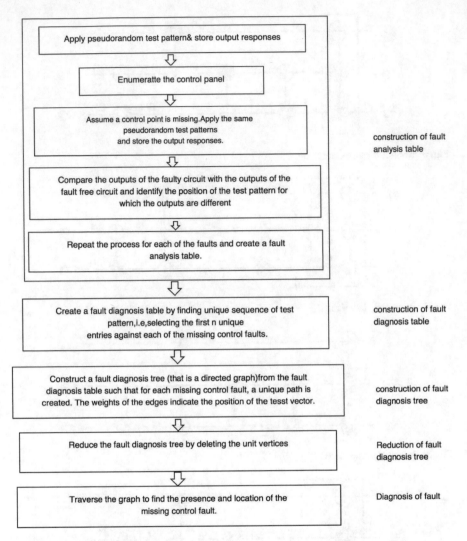

**Fig. 10** Block diagram of fault diagnosis method

## 4   Fault Testing in Reversible Circuits

In this section, we review some of the techniques to test the presence of the faults, introduced in the previous section, in a circuit. The testing can be done in four ways— (i) by constructing a Diagnostic Test Set (DTS), (ii) using voter insertion techniques, and (iii) using Binary Decision Diagrams (BDDs) (which is only applicable to SMGF and MMGF). We review each of these techniques individually in brief.

1. **Diagnostic Test Set (DTS) Method** [15]: In this method, initially test patterns from Automatic Test Pattern Generation (ATPG) are fed to the fault-free circuit and the outputs are recorded. When a single fault is encountered in the circuit, the same test patterns are applied on the circuit once more and the output is compared with the fault-free output. This allows to specify the location of the fault. If the circuit encounters more than one fault, then the same test is repeated with different test cases till all the faults are detected. For every positive result obtained, the test pattern is recorded in a fault diagnosis table and the corresponding tree structure is constructed. Tree structure has lesser search time compared to any linear data structure. Reference [4] hence, the choice of tree structure, reduces the time complexity of the whole process. The test set that is used to detect the faults is known as Diagnostic Test Set (DTS). A DTS of the reversible circuit shown in Fig. 10 is {000, 001, 110, 111}.

2. **Vector Insertion technique**: The voting technique for reversible logic was introduced based on majority multiplexing where a reversible majority gate named *MAJ* was introduced [30]. In a *MAJ*, the output is the majority of input bits. Hence, if a single input bit incurs fault, then the output may change. This is called *single-point failure*. Here, the two garbage outputs are used up for fault diagnosis method. There are two types of implementation: (a) Minimal Triplicated Voter(MTV) implementation and (b) Robust Triplicated Voter(RTV) implementation (Figs. 11 and 12).

   In case of MTV, there are four stages (i.e., four reversible gates). The analysis is made taking into consideration that a single fault results in three types of errors, viz., maskable error, recoverable error, and unrecoverable error. It still cannot properly detect the faults that have occurred in the input lines of voter. To increase the robustness, RTV was proposed. In RTV, the three copies of voted value are produced independently from the inputs in a direct manner. It reduced the chances of unrecoverable error by masking the single fault in the cost of increased area and delay. The diagnosis can be of single block level or multiple block level. For single block, say, if there are three copies of a sub-circuit with $k$ gates and a RTV voter, then the diagnosis is on the block of $(k + 8)$ gates. The garbage outputs only show the location of the fault on the data inputs used for fault detection in the sub-circuits. Each RTV produces two outputs.

   In case of multiple blocks with $n$ number of RTVs, we obtain $2n$ number of diagnosis outputs. The locations of the faults are identified through these outputs. Further, to reduce the time of monitoring, a special reversible circuit, called Diagnosis Collector (DC), was designed to collect all the diagnosis outputs as inputs and give a single output indicating the location of the faults in the circuit (Fig. 13).

3. **BDD Method**: BDD, which stands for Binary Decision Diagram, is only applicable for SMGF and MMGF. It starts with generation of test patterns. The main goal is to obtain a test set that will cover all possible faults in the circuit with minimum number of test patterns. The next and the most vital step is the dependency analysis, where the dependencies between all possible combinations of two fault gates are analyzed using the dependency analysis algorithm [26].

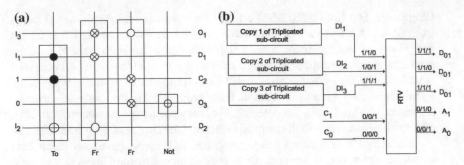

**Fig. 11  a** MTV and **b** RTV in TMR

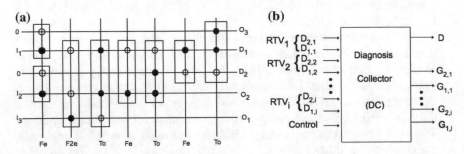

**Fig. 12  a** RTV and **b** Module collecting information of i RTVs

Consider a circuit with $m$ lines, then the BDD contains $2^m$ inputs and only one output. The same output can be represented by the minimum weighted path where the arc has a weight of 1 and other than that it has a weight of 0. Here, BDD containing SMGF test patterns are altered for all the MMGF resulting in a BDD with $n$ inputs (where $n$ is the number of lines) and $2 * {}^N C_2$ outputs (where $N$ is the number of gates).

**Universal Test Set**: All the techniques discussed above point us to the same direction—the construction of a unique test set that can detect all types of faults [20].

$$S_U = \begin{pmatrix} x_1 & x_2 \ldots x_n & c_x \\ 0 & 1 \ldots 1 & 1 \\ 1 & 0 \ldots 1 & 1 \\ \ldots & \ldots & \ldots \\ 1 & 1 \ldots 0 & 1 \\ 1 & 1 \ldots 1 & 0 \end{pmatrix}$$

An augmented CNOT gate is constructed as follows—Let a k-CNOT gate has an input sequence of $x_1, x_2, x_3, \ldots x_j, x_k, t$ and gives an output like $c_1, c_2, c_3, \ldots c_j, c_k$,

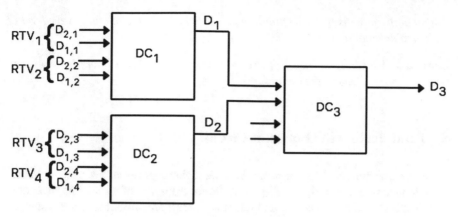

**Fig. 13** Cascaded DCs

$T, c_x$. This is done by adding an extra control line $(c_x)$ and repeating the same sequence $(x_1, x_2, x_3, \ldots x_j, x_k, t, c_x)$ of gates (k-CNOT gate) as embedded in the circuit, such that the extra control line can give a controlled output $(c_x)$ and has an output sequence of $c_1, c_2, c_3, \ldots c_j, c_k, T1, c_x$.

**Lemma 1** *We will get the same output at the target after augmentation, as the given input in t when $c_x = 1$. This can be simply shown as follows:*

$$
\begin{aligned}
T1 &= T \oplus (1, x_1, x_2, x_3 \ldots x_j \ldots x_k) \\
&= T \oplus (x_1, x_2, x_3 \ldots x_j \ldots x_k) \\
&= t \oplus (x_1, x_2, x_3 \ldots x_j \ldots x_k) \oplus (x_1, x_2, x_3 \ldots x_j \ldots x_k) \\
&= t
\end{aligned}
\tag{1}
$$

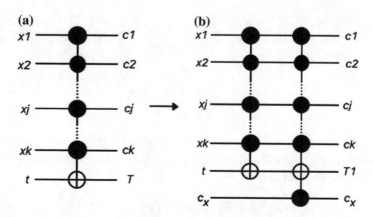

**Fig. 14** **a** A k-CNOT gate and **b** Augmented CNOT gate

*Note that the process of augmentation must be repeated for each and every level of logic implementation.*

**Lemma 2** *A test set matrix $S_U$ of size $(n + 1)$, shown in Fig. 14, will be enough to detect all missing gate faults of order $> 1$.*

## 5 Fault Testing in Quantum Circuit

Although quantum circuits are reversible, the technologies that are used to implement a quantum circuit [1, 16, 25] are different from that of a classical reversible circuit. Some faults in classical and quantum circuit are similar in nature, such as manufacturing fault, design error, physical defects, and probabilistic error. In addition to these, quantum circuits have some notable properties: (i) a quantum circuit may contain a non-detectable fault at the time of measurement and (ii) a quantum circuit may incorporate a fault that may be partially detected during measurements. This demands the need to characterize faults observed so far only in quantum circuits.

An ideal quantum circuit, i.e., a circuit without any fault, is termed *Gold Circuit* (GC). If, by executing a particular test set, one can determine all the faults present in a given circuit, then the test set is said to be a complete fault coverage. Fault localization [12] is a typical method that is inherited from classical fault model to rectify the possible faults. Due to the adaptation of the test set, this diagnostic method helps to locate the types of faults and their positions in the circuit. We provide a brief sketch of the fault models in quantum circuit.

**Pauli Fault Model**

Decoherence [17] in a qubit can be represented by the general rotation matrices

$$R_x(\theta) = \begin{pmatrix} cos(\theta/2) & -isin(\theta/2) \\ -isin(\theta/2) & cos(\theta/2) \end{pmatrix}, \quad R_y(\theta) = \begin{pmatrix} cos(\theta/2) & sin(\theta/2) \\ -sin(\theta/2) & cos(\theta/2) \end{pmatrix}$$

and

$$R_z(\phi) = \begin{pmatrix} e^{-\iota\phi/2} & 0 \\ 0 & e^{\iota\phi/2} \end{pmatrix}$$

Pauli matrices form the basis of $2 \times 2$ matrix space. Hence, any such arbitrary rotational matrix can be written as a linear combination of the Pauli matrices, whose spectral decompositions [17] are as

$$\sigma_x = |1\rangle \langle 0| + |0\rangle \langle 1| \quad \sigma_y = i|0\rangle \langle 1| - i|1\rangle \langle 0| \tag{2}$$

$$\sigma_z = |0\rangle \langle 0| - |1\rangle \langle 1| \quad I = |0\rangle \langle 0| + |1\rangle \langle 1| \tag{3}$$

Hence, any arbitrary fault can be written as an addition of one unwanted Pauli matrix $f$ in a quantum circuit (QC) at some error location $l$ with probability $p$. Some error correcting codes [6, 22, 24] have been proposed in the literature to correct a single-qubit Pauli error. Many faults in quantum circuits can be modeled using Pauli fault model, viz., depolarizing channel [11], phase dampening [8, 11], amplitude dampening [8], initialization inaccuracies [18], and measurement inaccuracies [2, 18]. Moreover, errors closer to the classical circuit, like pulse length error [8], off-resonance effects [11], and refocusing errors [8, 22], can also be described by the Pauli fault model. In paper [3], it is shown that $\sigma_x$ and $\sigma_z$ (refer Eq. 3) faults are reachable with any computational basis input state from $K$-CNOT gates. We check the percentage accuracy to detect fault in the circuit.

## 5.1 Initialization Faults [1, 3, 8]

A qubit $\cos\theta\,|0\rangle + \sin\theta\,|1\rangle$ can incur some rotational error $R_n(\delta)$, where $n \in \{x, y, z\}$ and the state changes to $\cos\theta\,|0\rangle + e^{i\epsilon}\sin\theta\,|1\rangle$. A qubit is said not to have initialization fault if it is not affected by such a rotational error. We have already stated before that any rotational error can be represented as a linear combination of the Pauli matrices. Hence, correcting only the Pauli matrices suffices. To correct $\sigma_x$ alone, a repetition code along with majority voting is sufficient. It was shown by Shor [22] that phase flip error is equivalent to bit flip in $\{|+\rangle, |-\rangle\}$ basis, where $|\pm\rangle = \frac{1}{\sqrt{2}}(|0\rangle \pm |1\rangle))$.

Now, we explain initialization fault with an example. Suppose the primary state is $|01c\rangle$, $c \in \{0, 1\}$ (Fig. 15), then by inverting the top qubit ($|c\rangle \Rightarrow |\bar{c}\rangle$), we get the final state as $\cos\theta\,|01c\rangle - i\sin\theta\,|11\bar{c}\rangle$. After an effect of Toffoli gate, the state is modified such that the probability of error becomes $(\sin\theta)^2$. Similarly, Fig. 15 also shows the situation where initialization fault impacted the center qubit. In order to

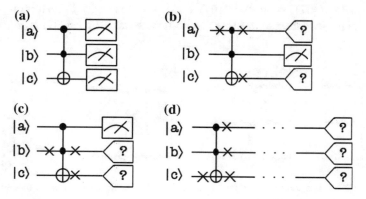

**Fig. 15** Initialization errors impacting a 2CN Gate: **a** correct circuit, **b–d** various initialization errors

**Table 1** Values for various faults in lost phase model

| term | initial | GC(a) | (b) | (c) | (d) |
|---|---|---|---|---|---|
| $|000\rangle$ | +1 | +1 | +1 | +1 | +1 |
| $|001\rangle$ | -1 | -1 | -1 | -1 | -1 |
| $|010\rangle$ | +1 | +1 | -1 | +1 | +1 |
| $|011\rangle$ | -1 | -1 | +1 | -1 | -1 |
| $|100\rangle$ | +1 | +1 | +1 | -1 | +1 |
| $|101\rangle$ | -1 | -1 | -1 | +1 | -1 |
| $|110\rangle$ | +1 | -1 | -1 | -1 | +1 |
| $|111\rangle$ | -1 | +1 | +1 | +1 | -1 |

detect initialization fault, the qubit must be measured once in $\sigma_z$ basis for bit flip and then again in $\sigma_x$ basis for phase flip.

## 5.2 Lost Phase Model [3, 8]

Lost phase model is basically a random phase error by allowing unwanted phase shift $\pm\epsilon$. A correct 3-CN gate may connect an input state $|+++\rangle\,|-\rangle$ to output state $(|000\rangle + |001\rangle + |010\rangle + |011\rangle + |100\rangle + |101\rangle + |110\rangle + |111\rangle)\,|-\rangle$ and here superposition term triggers the gate which will encounter a phase shift of $|n\rangle \Rightarrow e^{i\pi}\,|n\rangle$.

The impact of phase faults on the input state $|+++\rangle$ is described in Fig. 16 and Table 1. First column projects the phase of each term before being acted upon the circuit. The column GC (a) shows the correct relative phase of each term in entanglement. And another depicts phase changes due to the presence of fault.

If such a fault is present in the circuit of Fig. 16, then the output will be $(|000\rangle + |001\rangle + |010\rangle - |011\rangle + |100\rangle + |101\rangle + |110\rangle - |111\rangle)\,|-\rangle$. From this output, it is visible that relative phase shift occurs on both the states $|011\rangle$ and $|111\rangle$, since those

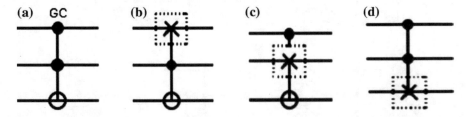

**Fig. 16** CCNOT gate and phase fault: **a** Gold circuit, **b** weak top control, **c** weak second control, **d** weak gate

activate the gate when the top control is broken. Another type of fault is phase dampening, i.e., a noise process altering relative phases between quantum states.

## 5.3   Measurement Fault Mode [2, 8]

Measurement faults occur due to the restriction in the responsiveness of a measurement apparatus. In addition to the Pauli fault model, a faulty measurement instrument is modeled as a probe that couples to a qubit and consistently returns a certain value. Here, each qubit must be measured in both logic-$|0\rangle$ and logic-$|1\rangle$ states (Fig. 17). In Fig. 18, we show the fault models described above and the test cases are described in Table 2.

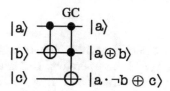

**Fig. 17**  Error free quantum circuit [3]

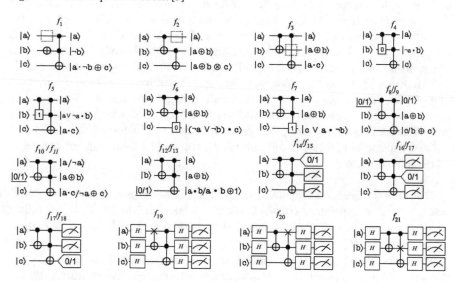

**Fig. 18**  Measurement Errors: Figs **a**, **b**, and **c** illustrate measurement faults that statistically favor logic 0. Figs **d**, **e**, and **f** contain measurement faults statistically favoring logic 1. [3]

**Table 2** Test patterns for faults shown in Fig. 18

| abc | GC | $f_1$ | $f_2$ | $f_3$ | $f_4$ | $f_5$ | $f_6$ | $f_7$ |
|-----|-----|-------|-------|-------|-------|-------|-------|-------|
| 000 | 000 | 010 | 000 | 000 | 000 | 000 | 000 | 000 |
| 001 | 001 | 011 | 001 | 001 | 001 | 001 | 001 | 001 |
| 010 | 010 | 000 | 011 | 010 | 010 | 010 | 010 | 010 |
| 011 | 011 | 001 | 010 | 011 | 011 | 011 | 011 | 011 |
| 100 | 111 | 111 | 111 | 101 | 100 | 111 | 110 | 110 |
| 101 | 110 | 110 | 110 | 100 | 101 | 110 | 111 | 111 |
| 110 | 100 | 100 | 100 | 111 | 100 | 111 | 100 | 101 |
| 111 | 101 | 101 | 101 | 110 | 101 | 110 | 100 | 101 |

# 6   Advantages and Disadvantages of Fault Testing in Quantum Circuit

Every quantum operator is associated with a unitary matrix [17]. However, some of the quantum operators, viz., Pauli operators, Hadamard, and CNOT, are self-adjoint. A matrix $A$ is said to be self-adjoint if $A = A^\dagger$. Due to some error in technology, a particular gate $G$ may have been applied multiple times instead of the desired single occurrence of the gate. However, if $G$ is a self-adjoint operator, then $G.G^\dagger = G.G = I$. Hence, if a self-adjoint operator $G$ is applied $n$ times due to a fault, two scenarios can occur—(i) if $n$ is even, then $G.G \ldots G = I$ and (ii) if $n$ is odd, then $G.G \ldots G = G$. So, if the gate is applied odd number of times, then it does not account for any fault and if it is applied even number of times, then it is similar to SMGF. So, for self-adjoint operators in a quantum circuit, multiple occurrences of a gate does not lead to any new fault model. A test pattern that can identify single missing gate fault will also be able to identify fault due to multiple occurrences of the gate. This is indeed an advantage of quantum circuits over classical ones. However, not every quantum operator is self-adjoint. Hence, this advantage is not general to all quantum operators.

However, the question remains whether it is possible to generate test patterns to detect SMGF in a quantum circuit. Consider a simple circuit consisting of only a single Hadamard gate. If the initial state is $|0\rangle$, then after the Hadamard operation, the final state is $\frac{1}{\sqrt{2}}(|0\rangle + |1\rangle)$.

If the Hadamard gate is not applied, then the input state remains unchanged. But since $|0\rangle$ and $\frac{1}{\sqrt{2}}(|0\rangle + |1\rangle)$ are not orthogonal, no projective measurement can reliably distinguish between these two states [17]. Hence, it does not seem possible to detect whether the Hadamard gate has been applied or not. Further research is necessary to investigate whether the inclusion of ancilla qubits can be of some aid or whether it is possible to apply suitable POVM measurement for fault detection in quantum circuit.

# 7   Conclusion

In this paper, we have discussed fault testing in reversible and quantum circuits. We have given a detailed description of fault models in classical and quantum circuits and provided remedies for their detection. We have also shown with an example that since it is not possible to distinguish two non-orthogonal quantum states, it may not be possible to detect certain faults in a quantum circuit. While fault testing has been studied largely in classical circuits, research on quantum circuits is still mostly limited to error correction. It will be worthwhile to pursue more studies on quantum fault models, which are likely to depend on and vary with the various technologies that are used presently to implement a quantum computer.

# References

1. Barrett, M.D., Schätz, T., Chiaverini, J., Leibfried, D., Britton, J., Itano, W.M., Jost, J.D., Knill, E., Langer, C., Ozeri, R., et al.: Quantum information processing with trapped ions. In: AIP Conference Proceedings, vol. 770, pp. 350–358. AIP (2005)
2. Bennett, C.H.: Logical reversibility of computation. IBM J. Res. Dev. **17**(6), 525–532 (1973)
3. Biamonte, J.D., Allen, J.S., Perkowski, M.A.: Fault models for quantum mechanical switching networks. J. Electron. Test. **26**(5), 499–511 (2010)
4. Cormen, T.H.: Introduction to Algorithms. MIT press, Cambridge (2009)
5. Feynman, R.P.: Quantum mechanical computers. Found. Phys. **16**(6), 507–531 (1986)
6. Gottesman, D.: Stabilizer codes and quantum error correction (1997). arXiv preprint quant-ph/9705052
7. Grover, L.K.: A fast quantum mechanical algorithm for database search. In: Proceedings of the Twenty-eighth Annual ACM Symposium on Theory of Computing, pp. 212–219. ACM (1996)
8. Knill, E., Laflamme, R., Ashikhmin, A., Barnum, H., Viola, L., Zurek, W.H.: Introduction to quantum error correction (2002). arXiv preprint quant-ph/0207170
9. Knill, E., Laflamme, R., Viola, L.: Theory of quantum error correction for general noise. Phys. Rev. Lett. **84**(11), 2525 (2000)
10. Landauer, R.: Irreversibility and heat generation in the computing process. IBM J. Res. Dev. **5**(3), 183–191 (1961)
11. Lee, S., Lee, S.-J., Kim, T., Lee, J.-S., Biamonte, J., Perkowski, M.: The cost of quantum gate primitives. J. Mult. Valued Log. Soft Comput. 12 (2006)
12. Li, C., Liu, L., Pang, X.: A dynamic probability fault localization algorithm using digraph. In: 2009 Fifth International Conference on Natural Computation, August 2009, vol. 6, pp. 187–191 (2009)
13. Majumdar, R., Basu, S., Mukhopadhyay, P., Sur-Kolay, S.: Error tracing in linear and concatenated quantum circuits (2016). arXiv preprint arXiv:1612.08044
14. Majumdar, R., Saini, S.: A novel design of reversible 2: 4 decoder. In: 2015 International Conference on Signal Processing and Communication (ICSC), pp. 324–327. IEEE (2015)
15. Mondal, B., Das, P., Pradyut, S., Chakraborty, S.: A comprehensive fault diagnosis technique for reversible logic circuits. Comp. Electr. Eng. **40**(7), 2259–2272 (2014)
16. Munro, W.J., Nemoto, K., Spiller, T.P., Barrett, S.D., Kok, P., Beausoleil, R.G.: Efficient optical quantum information processing. J. Opt. B: Quantum Semiclassical Opt. **7**(7), S135 (2005)
17. Nielsen, M.A., Chuang, I.L.: Quantum Computation and Quantum Information. Cambridge university press, Cambridge (2010)
18. Obenland, K.M., Despain, A.M., Turchette, T.Q.A., Hood, C.J., Lange, W., Mabuchi, H., Kimble, H.J., et al.: Impact of errors on a quantum computer architecture (1996)

19. Peres, Asher: Reversible logic and quantum computers. Phys. Rev. A **32**(6), 3266 (1985)
20. Rahaman, H., Kole, D.K., Das, D.K., Bhattacharya, B.B.: On the detection of missing-gate faults in reversible circuits by a universal test set. In: 21st International Conference on VLSI Design. VLSID 2008. pp. 163–168. IEEE (2008)
21. Saligram, R., Hegde, S.S., Kulkarni, S.A., Bhagyalakshmi, H.R., Venkatesha, M.K.: Design of fault tolerant reversible multiplexer based multi-boolean function generator using parity preserving gates. Int. J. Comput. Appl. **66**(19) (2013)
22. Shor, P.W.: Scheme for reducing decoherence in quantum computer memory. Phys. Rev. A **52**(4), R2493 (1995)
23. Shor, P.W.: Polynomial-time algorithms for prime factorization and discrete logarithms on a quantum computer. SIAM Rev. **41**(2), 303–332 (1999)
24. Steane, A.M.: Error correcting codes in quantum theory. Phys. Rev. Lett. **77**(5), 793 (1996)
25. Strauch, F.W., Johnson, P.R., Dragt, A.J., Lobb, C.J., Anderson, J.R., Wellstood, F.C.: Quantum logic gates for coupled superconducting phase qubits. Physical Rev. Lett. **91**(16), 167005 (2003)
26. Surhonne, A.P., Chattopadhyay, A., Wille, R.: Automatic test pattern generation for multiple missing gate faults in reversible circuits. In: International Conference on Reversible Computation, pp. 176–182. Springer (2017)
27. Thapliyal, H., Ranganathan, N.: Design of reversible sequential circuits optimizing quantum cost, delay, and garbage outputs. ACM J. Emerg. Technol. Comput. Syst. (JETC) **6**(4), 14 (2010)
28. Tommaso, T.: Reversible computing. In: Automata, Languages and Programming, pp. 632–644 (1980)
29. Wille, R., Drechsler, R.: Bdd-based synthesis of reversible logic for large functions. In: Proceedings of the 46th Annual Design Automation Conference, pp. 270–275. ACM (2009)
30. Zamani, M., Farazmand, N., Tahoori, M.B.: Fault masking and diagnosis in reversible circuits. In: 16th IEEE European Test Symposium (ETS), pp. 69–74. IEEE (2011)

# Part II
# Distributed Systems and Security

# TH-LEACH: Threshold Value and Heterogeneous Nodes-Based Energy-Efficient LEACH Protocol

**Pratima Sarkar and Chinmoy Kar**

**Abstract** Sensor nodes are used to measure ambient of environment. Sensor network can be defined as a collection of sensor nodes, which senses the environment and sends information to the base station. These types of networks are facing problems related to energy dissemination. LEACH (Low Energy Adaptive Cluster Hierarchy) protocol is one of the most suitable protocols used for communication in a sensor network. LEACH protocol is a cluster-based protocol, in which each cluster consists of multiple nodes and one cluster head node. The cluster head aggregates all the information from other nodes in the cluster and conveys it to the base station. In LEACH protocol, cluster head selection is performed in each round at the cost of some amount of energy. In our proposed solution, the energy consumption for electing cluster head in every round is circumvent. This can be achieved by threshold value-based cluster formation. The proposed method reduces the overhead of forming cluster in every round, which helps in reduction of energy consumption. As a result of the work, the performance of proposed solution is compared among a variety of LEACH protocols with respect to lifetime of the network.

**Keywords** LEACH · Energy efficient · Wireless sensor network

## 1 Introduction

Wireless Sensor Network (WSN) consists of sensor nodes that collecting the information about the environment and transmits it to the base station. In this type of network, sensor nodes have limited amount of energy. WSN network is very useful to serve various parameters in remote and hostile regions as in detecting attacks, monitoring enemies' movement, etc. Routing protocols have very important role in

P. Sarkar (✉) · C. Kar (✉)
Sikkim Manipal Institute of Technology, Sikkim, India
e-mail: psmoon2@gmail.com

C. Kar
e-mail: info.chinmoy@gmail.com

© Springer Nature Singapore Pte Ltd. 2018
S. K. Das and N. Chaki (eds.), *Algorithms and Applications*, Smart Innovation,
Systems and Technologies 88, https://doi.org/10.1007/978-981-10-8102-6_3

this type of network. LEACH protocol is a cluster-based protocol that can be used in WSN. The protocol is designed to preserve the maximum amount of energy. This protocol comprises two phases—setup phase and steady-state phase.

In LEACH protocol, setup phase is required in each round [1]. Selection of cluster head is based on threshold value T(n).

$$T(n) = p/\left[1 - p \times r\{mod(1/p)\}\right]$$  (1)

In Eq. 1, p is percentage of nodes which are becomes a cluster head in each round, and r is number of rounds. The nodes which are not selected in the previous round will get chance to be cluster head. In each round of LEACH, all the nodes are assigned with a random value between 0 and 1. Any node having the random value less than T(n) will be elected as a cluster head. After the election of cluster head, it broadcasts message to other nodes. Other nodes join the cluster head depending upon signal strength send by different cluster heads [2]. TDMA scheduling is used for data transmission. The cluster heads make a TDMA schedule and send it to the all the other nodes available in the cluster [1]. As per the given TDMA schedule, other nodes send data to the cluster head. These all collected data aggregated by cluster head and sent it to the base station. The non-cluster head nodes turn on its radio component only during data transmission [2].

In the steady phase of LEACH protocol, it starts data transmission. Here it is assumed that all the nodes are ready to send information in its given slot [1]. After receiving all the information, the cluster heads aggregate and send the information to the base station.

## 2 Literature Survey

There are so many exiting parameters for measuring the efficiency of any routing protocol but out of them most important parameter for measuring the performance of wireless sensor network is lifetime of the network that is directly proportional to the energy available to the network.

In [3], LEACH-DT focuses on selection of cluster head that depends upon the residual energy of each node and distance from the base station. The solution proposed in [3], the cluster head will transfer responsibilities to a node with the highest residual energy. In the result section, the modified LEACH-DT and LEACH-DT protocols are compared with respect to remaining energy available in the whole network and lifetime of the network.

K-LEACH [4] protocol concentrates on uniform clustering of nodes and one of the best selections of cluster heads. K-medoids algorithm is used for uniform selection of head. Euclidian distance is also considered to make selection of cluster head closest to the previous cluster head. After first round, cluster heads are elected depending upon the position of previous cluster head and so on. Here the proposed work compares the performance of LEACH protocol and K-LEACH protocol with respect to energy retention and number of live nodes per round.

In paper [2], a very clear impression of overall energy required for setup phase and steady-state phase is given. It includes the idea to calculate the length of each round. Round is defined as when every node in a cluster completes one-time data transfer to the cluster head for existing TDMA schedule. The performance of LEACH protocol in wireless sensor networks is analyzed with respect to lifetime and throughput. Loss of energy by cluster head in each steady state gives an idea for calculating threshold value this work.

IB-LEACH [5] is a protocol that supports heterogeneous energy level for few nodes and decreases the failure probability of nodes. As mentioned earlier, the IB-LEACH protocol consists of two phases: setup and steady-state phases. Setup phase mainly divided into various components like gateway selection, cluster formation, and so on. The probability of electing cluster gateway depends upon threshold value. In a cluster, there are two types of nodes—normal and advanced nodes. These nodes are assigned with different probability values. Depending upon the probability values, nodes are elected as cluster heads. At the end of this paper, IB-LEACH is compared with LEACH protocol with respect to number of alive nodes available in each round.

## 3 Proposed Solution

Most of the existing works concentrate on best cluster head selection in each round, which leads to some amount of energy consumption. In this work, we have tried to avoid cluster formation in each round so that we can save energy required for making cluster or setup phase. Heterogeneous energy level is assigned to some of the sensor nodes, and it helps in preserving energy. This can be achieved by avoiding cluster formation as much as possible throughout network's lifetime. There are two phases of LEACH protocol—setup phase and steady-state phase. In this proposed solution, changes are made in setup phase only.

The proposed solution is as follows:

1. Probability $p = 0.1$ implies 10% of nodes in complete network are selected as cluster head. Depending on the value of p, i.e., 10% node of the whole network is made as heterogeneous node. These nodes encompass different levels of energy than normal nodes.
2. Energy level of heterogeneous node is dependent on initial energy level of normal nodes, i.e., $E_{in}$. $E_{ch} = 2*E_{in}$.
3. Positions of the heterogeneous nodes are placed in such a way that it covers the whole area of a network and elected as cluster head in first round. Normal nodes are placed with some random value. Coordinates for heterogeneous nodes are calculated by

$$x(i) = (s/2*N*p) + (i/p) \text{ and } y(i) = (s/2*N*p) + (i/p) \quad (2)$$

**Table 1** Notation details for above algorithm

| Notation used | Explanation |
| --- | --- |
| R | Round number |
| N | Number of alive nodes |
| E(i) | Energy level of $i_{th}$ node |
| CH | Cluster head |
| CH.E | Energy level of cluster head |

where $(x(i),y(i))$ represents coordinates of $i_{th}$ heterogeneous node, p is percentage of nodes to be cluster head, N is number of nodes, and s is size of network.

4. Cluster head selection is dependent on energy level of a node. If the energy of a node is greater than the average energy of all other nodes present in the network then that node is selected as cluster head. Cluster head election will be organized if and only if energy level of the cluster head is less than T(n), where n is number of nodes available in a cluster. This approach minimizes the overhead of cluster head selection in each round. The value of T(n) is decided depending upon energy required to serve one round in steady-state phase by cluster head. The total energy required to serve a round in steady state by cluster head and communication energy is same as [2].

T(n) = Total energy required to serve a round in steady state a for cluster head.

T(n) = Energy required for receiving (n−1) frames + Data aggregation for n nodes including cluster head + Transmission of aggregated data to base station (Eq. 3, Table 1).

$$T(n) = (n - 1) * L*E_{elec} + n*L*E_{DA} + (L*E_{elec} + L* E_{mp} * d_{to\ Bs}^4) \qquad (3)$$

n       = Number of nodes in a cluster.
L       = Number of bits available in a frame.
$E_{elec}$  = Expenditure of energy for receiving or transmitting one bit.
EDA   = Data aggregation energy.
$E_{mp}$   = 0.0013 PJ/bit/m4.
d       = Distance between cluster head and base station.

### Algorithm for cluster head selection:

Step 1: Initialize r=1
Step 2: Initialize j= N*.01
Step 3: count=0;
Step 4: While i<N

Step 4.1: if $E(i) > \sum_{i=1}^{i=N} \frac{E(i)}{N}$ then  // Identifying higher energy nodes.
Step 4.1.1: count=count+1;
Step 4.1.2 : if  count<=j then
Step 4.1.2.1:  CH(count)<-N(i) // Node N(i) selected as cluster head.

                                        end if
                              end if
          Step 5: r=r+1 and send TDMA schedule;
          Step 6: When r>1
                    Step 6.1: for i=1 to N
                              Step 6.1.1: if CH.E>T(n) //Energy of cluster head greater than
                                        Threshold value
                                        Step 6.1.1.1: r=r+1 send TDMA schedule;
          Step 6.1.2: else // Energy of cluster head below Threshold value
          Step 6.1.2.1: N=numbers of alive node
                    Step 6.1.2.2: goto Step 2.

5.  The amount of energy saved by in case of "the energy level of cluster head more
    than threshold value" calculated as follows:

    Energy saving by non-cluster head nodes in each round is

          $E_{node}$ = Energy required to (receive broadcast + send join request)

          $= (p*E_{elec}) + (p*E_{elec} + p*E_{fs}*d^2_{toCH})$                    (4)

    Energy saving by cluster head nodes in each round is

          $E_{CH}$ = Energy required to (broadcast + receiving join request)

          $= (p*E_{elec} + p*E_{fs}*d^2_{toNode}) + p*E_{elec}*(n - 1)$           (5)

$d_{toCH}$     = Distance between the node to cluster head.
$d_{toNode}$   = Distance between the cluster head to a particular node.
n              = Numbers of nodes available in a cluster.
$E_{fs}$       = Amplification energy.

6.  After selection, cluster head will send signals to other clusters for joining.
    Depending upon signal strength received by different cluster heads, other nodes
    join them. These are the modification made in setup phase but none in the steady-
    state phase.


# 4  Simulation Environment

This work is simulated on MATLAB simulator tool. In the region on $400 \times 400$ m$^2$,
normal numbers of nodes are placed randomly. Heterogeneous nodes are placed in
such a way so that nodes are elected as cluster head.

    Coordinates for heterogeneous nodes are calculated by $x(i) = (s/2*N*p) + (i/p)$ and
$y(i) = (s/2*N*p) + (i/p)$ (as explained in solution strategy). Heterogeneous nodes are
initialized with more amount of energy than normal nodes. There are four scenarios

**Table 2** Simulation parameters

| Network parameters | Values |
|---|---|
| Network size | $400 \times 400$ m$^2$ |
| No. of nodes | 100, 200, 300, 400 |
| Initial energy of normal sensor nodes | 0.5 J |
| Initial energy of heterogeneous sensor nodes | 1 J |
| Packet size | 4000 bits |
| Transceiver idle state energy consumption | 50 nJ/bits |
| Amplification energy | Efs = 10 pJ/bit/m3<br>Emp = 0.0013 pJ/bit/m3 |
| Data aggregation energy | EDA = 50 nJ/bit |

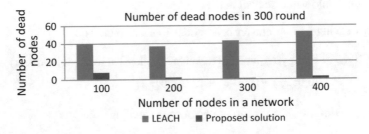

**Fig. 1** Comparison graph for numbers of dead nodes in 300 rounds

considered for 100, 200, 300, and 400 nodes, and in each of the simulation, lifetime of the LEACH protocol is compared with the proposed solution (Table 2).

## 5 Result and Discussion

In this work, LEACH protocol is compared with the proposed solution based on the following parameters: improvement in lifetime of the network, first dead node, and number of dead nodes for 300 rounds.

First dead node means that the number of round required by a node to die. Lifetime of any network determines the minimum number of round in which all nodes of the network are dead.

Figure 1 shows number of dead nodes in 300 round. In the proposed solution, number of dead nodes is much less than LEACH protocol for all the different scenarios. The reason behind this is that the selected cluster heads are heterogeneous nodes which have more amount of energy than other nodes, and they will be able to serve more numbers of rounds.

The graph in Fig. 2 considers four different scenarios with 100, 200, 300, and 400 nodes. Here y-axis denotes a number of rounds that are used to represent in which

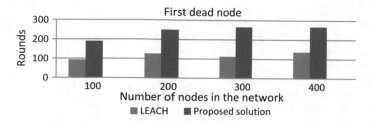

**Fig. 2** Comparison graph for first dead node in a network

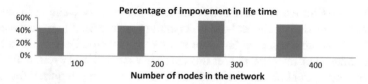

**Fig. 3** Percentage of improvement in lifetime of network with respect to LEACH protocol

round first node of the network dies. The above graph clearly shows the proposed solution giving a better result than LEACH protocol. In LEACH protocol, selection is done randomly but in this work, it is done on the basis of maximum residual energy. In first round, random nodes are placed such a way that they are selected as cluster head. Cluster selection is not for every round, and hence, the energy consumption for cluster formation is also reduced.

Figure 3 considers four different situations where % of improvement in lifetime is plotted with increasing numbers of nodes.

$$\% \text{ of improvement in life time } = \frac{\text{Lifetime in LEACH Protocol}}{\text{Life time in proposed solution}} * 100 \quad (6)$$

Figure 3 shows that in all scenarios, the proposed solution gives a better result than LEACH protocol. The performance of the network improves because of proper selection of cluster head in the network.

Table 3 shows that the improvement in first node dies with respect to LEACH outweigh the improvement in last node dies with respect to LEACH. Our method shows better result with respect to other standard methods when second parameter considered.

## 6 Conclusion

In this work, a novel approach is proposed for increasing lifetime of the network. The proposed protocol for wireless sensor network is a modification of LEACH protocol. Figure 3 shows that lifetime of the network has improved by more than

**Table 3** Comparison with existing art-of-studies

| Parameter | K-LEACH (%) | IB-LEACH (%) | Proposed work (%) |
|---|---|---|---|
| Improvement in first node dies with respect to LEACH | 41.8 | 61.6 | 50 |
| Improvement in last node dies with respect to LEACH | 0 | 7.3 | 43 |

40%. Lifetime of the network is directly proportional to energy available in the nodes. In this work, selection of cluster head is depending on maximum residual energy so cluster head can able to serve for more number of rounds. Preserving energy of nodes is achieved by avoiding cluster formation in each and every round. Formation of cluster is dependent on the threshold value. It saves energy required for creating a new cluster and also increases the lifetime of the network.

## 7 Future Scope

This paper shows promising result to save energy of a network and also to increase the lifetime of network. Security issues related to LEACH protocol will be addressed further, which may improve the accuracy and effectiveness of the work.

## References

1. Heinzelman, W.R., Chandrakasan, A., Balakrishnan, H.: Energy-efficient communication protocol for wireless microsensor networks. In: Proceedings of the Hawaii International Conference on System Sciences, pp. 2–10. IEEE, Maui, Hawaii (2000)
2. Li, Y., Nan, Y., Zhang, W., Zhao, W., You, X., Daneshmand, M.: Enhancing the performance of LEACH protocol in wireless sensor networks. In: IEEE Infocom 2011
3. Gupta, V., Pandey, R.: Modified LEACH-dt algorithm with hierarchical extension for wireless sensor networks. Int. J. Comput. Netw. Inf. Sec. **2**, 32–40 (2016)
4. Bakaraniya, Parul, Mehta, Sheetal: K-LEACH: an improved LEACH protocol for life-time improvement in WSN. Int. J. Eng. Trends Technol. (IJETT) **4**(5), 1521–1526 (2013)
5. Said, B.A., Abderrahim, E.A., Hssane, B., Hasanoui, M.L.: Improved and balanced LEACH for heterogeneous wireless sensor networks. Int. J. Comput. Sci. Eng. **2**(8), 2633–2640 (2010)
6. Kumar, N., Kaur, J.: Improved LEACH (I-LEACH) protocol for wireless sensor networks. IEEE (2011)
7. Sivakumar, B., Sowmya, B.: An energy efficient clustering with delay reduction in data gathering (EE-CDRDG) using mobile sensor node. Wireless Pers. Commun. **90**(2), P793–782 (2016)
8. Chen, J.: Improvement of LEACH routing algorithm based on use of balanced energy in wireless sensor. In: International Conference on Intelligent Computing ICIC 2011, pp. 71–76. Advanced Intelligent Computing (2011)
9. Ran, G., Zhang, H., Gong, S.: Improving on LEACH protocol of wireless sensor networks using fuzzy logic. J. Inf. Comput. Sci. **7**(5), 67–77 (2010)

10. Bajelan, M., Bakhshi H.: An adaptive LEACH-based clustering algorithm for wireless sensor networks. J. Commun. Eng. 2(4), (2013)
11. Xinhua, W., Sheng, W.: Performance comparison of LEACH and LEACH-C protocols by NS2. In: 2010 Ninth International Symposium on Distributed Computing and Applications to Business, Engineering and Science, pp. 254–258 (2010)
12. So-In, C., Udompongsuk, K., Phudphut, C., Rujirakul, K., Khunboa, C.: Performance evaluation of leach on cluster head selection techniques in wireless sensor networks. In: The 9th International Conference on Computing and Information Technology (IC2IT2013). Advances in Intelligent Systems and Computing, vol. 209. Springer, Berlin, Heidelberg (2013)
13. Mehta, R., Pandey, A., Kapadia, P.: Reforming clusters using C-LEACH in wireless sensor networks. In: International Conference on Computer Communication and Informatics, pp. 1–4. IEEE Press, India (2012)
14. Pantazis, N.A., Nikolidakis, S.A., Vergados, D.D.: Energy-efficient routing protocols in wireless sensor networks: a survey. IEEE Commun. Surv. Tutor. 15(2), 1–41 (2012)
15. Mahapatra, R.P., Yadav, R.K.: Descendant of LEACH based routing protocols in wireless sensor networks. Procedia Compu. Sci. 57, 1005–1014 (2015)

# A Novel Symmetric Algorithm for Process Synchronization in Distributed Systems

Sourasekhar Banerjee, Prasita Mukherjee, Sukhendu Kanrar and Nabendu Chaki

**Abstract** While symmetric mutual exclusion algorithms are easy to implement, message complexity per critical section (CS) access to such approaches in a distributed system is typically high. There exist works that handle this issue but to a limited extent. In this paper, we propose a new symmetric algorithm for mutual exclusion. The proposed approach is essentially a prioritized version of the well-known Ricart–Agrawala algorithm for mutual exclusion in distributed systems. The solution proposed uses one or more priority levels, such that different participating processes are placed at different priority levels depending on the initial priority of the processes. The proposed algorithm maintains safety, liveness, and fairness properties toward implementation in a distributed system.

**Keywords** Mutual exclusion · Critical section · Symmetric · Distributed system
Priority · Permission based · Progress condition

S. Banerjee (✉) · N. Chaki
Department of Computer Science & Engineering, University of Calcutta, JD-2, Sector – III, Salt Lake City, Kolkata 700106, India
e-mail: sourasb05@gmail.com

N. Chaki
e-mail: nabendu@ieee.org

P. Mukherjee
Computer Science and Engineering, core 2, Academic complex, IIT Guwahati, Guwahati 781039, Assam, India
e-mail: prasitamukherjeemou@gmail.com

S. Kanrar
Department of Computer Science, Narasinha Dutt College, Howrah 711101, India
e-mail: sukhen2013@gmail.com

© Springer Nature Singapore Pte Ltd. 2018
S. K. Das and N. Chaki (eds.), *Algorithms and Applications*, Smart Innovation, Systems and Technologies 88, https://doi.org/10.1007/978-981-10-8102-6_4

51

# 1   Motivation and Problem Definition

In a distributed system, mutual exclusion (ME) means, at any given time only one process is allowed to access critical section (CS). The design of distributed mutual exclusion algorithm (DMEA) is not easy because these algorithms have to deal with unpredictable message delays and have incomplete knowledge of the system state. There are three basic approaches to achieve mutual exclusion in a distributed environment, which are symmetric non-token-based approach, token-based approach, and quorum-based approach. In non-token-based approach, two or more successive rounds of message exchanges between processes are required to decide which process will enter critical section next. In the token-based approach, a process gets permission to enter CS if it holds the token and releases it when the execution of CS is over. In the quorum-based approach, each process needs to seek only permission from the subset of other processes to execute the CS. In this manuscript, we will propose a new algorithm for non-token-based symmetric mutual exclusion which is an extension of Ricart–Agrawala (RA) algorithm. In Sect. 2, we have given a brief description on permission-based DMEA. In Sect. 3, we have discussed our proposed approach. The performance of the proposed approach has been given in Sect. 4. Simulation results establish the superiority of the performance of the proposed algorithm as compared to RA which is given in Sect. 5.

# 2   Review on Permission-Based ME Algorithms

We are working on permission-based algorithms for mutual exclusion for critical section access in distributed operating system. Ricart and Agrawala (RA) have proposed a symmetric non-token-based algorithm [1] which is an optimized version of Lamport's algorithm [2]. Lodha–Kshemkalyani's [3] algorithm is a modification of RA algorithm. In token-based approach, Kanrar–Chaki algorithm [4] is an extended version of Raymond's algorithm [5]. Lejeune [6, 7] has proposed an extended version of Kanrar–Chaki algorithm. The goal of their algorithm is to minimize priority violation without introducing starvation. A. Swaroop and A. Kumar Singh have also proposed algorithms on distributed mutual exclusion in groups [8, 9], which are also token based. In [10], the proposed algorithm is token permission based on prioritized groups. Maekawa's mutual exclusion algorithm [11] was the first quorum-based algorithm. Here a requesting process needs permission only from a subset of total processes. Atreya et al. [12] is another quorum-based group mutual exclusion algorithm which contains dynamic groups. Progress condition may be violated in voting algorithms when no single process gets the majority of votes. In [13], the proposed algorithm is a token-based mutual exclusion algorithm for cyclic or acyclic directed graph topology. Kanrar–Chaki [14] has proposed a new two-phase, hybrid ME algorithm that works even when majority consensus cannot be reached. Singhal's dynamic information-structure algorithm [15] is an adaptive mutual exclusion algorithm based on some observation. Naimi and Thiare [16] have proposed an approach of mutual exclusion in distributed systems using causal ordering. Here, they implemented the causal ordering in Suzuki–Kazami's [17] token-based algorithm that realized mutual exclusion among N processes. An energy-efficient algorithm for distributed mutual exclusion in mobile ad hoc networks has been given in [18].

## 3 The Proposed Symmetric Algorithm

The proposed approach is an extension of RA algorithm. In our algorithm, we are assigning several priority levels. Process in higher priority level will be served first and gets access to enter the critical section. Convention followed here is lesser the number, higher the priority. Each process has *external* as well as *internal* priority. We assume that priority is a positive real number. External priority is an integral part and internal priority is the fractional part of that number. We assign priority in many ways: Suppose a process is executing very frequently then its priority will be higher or priority of a process depends on the quality of jobs. According to external priorities, we are dividing the graph (Fig. 1) into subgraphs (Fig. 2). Processes with the same external priority will be placed in the same subgraph. In a subgraph, two or more processes with the same priority will get served in FCFS order. Using our approach we are applying RA only in one level at a time. If we have any process in the first level which wants to get into the CS, a request message will be sent to all other processes in that level. We use a global list to maintain the processes (with priority), which are interested in getting into the CS. The concept of aging has also been implemented by increasing the priority of the waiting processes by a random factor.

**Fig. 1** K16 graph

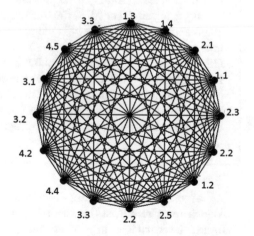

**Fig. 2** K16 graph divided
into 4 complete subgraphs

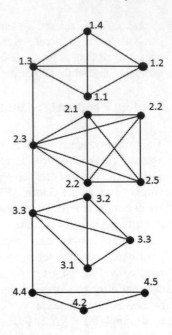

**Data Structure and Proposed Algorithm (PA)**

---

**Adjacency list structure**: This structure contains the participating processes in a distributed system. These processes may or may not wish to get into CS. By using adjacency list, the time complexity is $O(N + E)$, and the space complexity is $O(N)$ where N is the total participating processes and E is the total number of edges.

**Global List (GL)**: It is used to maintain requesting processes.

**Local List (LL)**: It is used to implement RA algorithm at each level.

**NODE**: The Process that contains burst time, priority, request flag (1 when it wishes to get in CS and 0 when it is not interested in CS immediately), and links on the left right, up and down directions. Initially, all these links are set to NULL.

---

**Algorithm_1: Inserting node into adjacency list**
**Algorithm: CreateList (int record[ ], int priority[ ], NODE\* head)**
**Input: 1. record** is an array of type integer which contains the details of each node.
**2. Priority** is an array of type double which contains the priority of each prcess.
**3. head** is a node pointing to the root of the adjacency list.
**Output:** adjacency list of all process in the distributed system.
**Return type:** NODE

**Start:**
Total_num_process = Total number of the process will be present in the list.
**do** {
Create a new NODE node1 values taken from record [ ] and priority [ ]
**if** (head==NULL)
head = node1;
**else** {
   external_priority_of_node1= floor (node1-> priority)
   external_priority_of_head = floor (head->priority)

**Case1: node1's external priority is same as head's external priority**
   **if** (external_priority_of_node1 ==external_priority_of_head){
node1 and head have the same priority level. Add node1 at the end of that
level.}

**Case2: node1's external priority is lesser than head's external priority.**
   **else if**(external_priority_of_node1 >external_priority_of_head) {
               // convention used here is lesser the number, higher in priority
     1. Move head one step down and check whether external priority of
       node1 is same with head.
    2. Continue this process until the external priority of node1 is same with the
head.
      3. if priority level has found add this node at the end of this level.
      else create a new level with node1 }

**Case3: node1's priority is greater than head's priority**
   **else** {
   Priority of node1 is higher than head so add new priority level above head's
     priority level and make head=node1}
   }
   process_placed_in_list ++;
   } **while** (process_placed_in_list<Total_num_process);
   Return head;
**End**


**Algorithm_2: Service of requested processes and aging of low priority processes.**
**Algorithm:** service (Node* head)
**Input:** head is a node pointing to root of the adjacency list.
**Output: 1.** Total number of message exchanges
     **2.** Turnaround time and waiting time of each Critical section executing
       process
**Start:**
   GL is a list to keep all requesting nodes;
   LL is a list for implementing RA algorithm at each level
   R=head of the adjacency list.

**do**{

Traverse the list corresponding to R.

**do**{

  **if** (Pr$_i$ == found) // Pr$_i$, is a requesting process at level i.

  {

    1.   Pr$_i$ sends a request message to all nodes in its level.

    2.   If any process who is not interested in getting into CS will send a "Go Ahead" message to Pr$_i$.

    3.   If any "interested" process with internal priority higher than the requesting process will send message "Stop" to Pr$_i$ and processes with the highest internal priority that wish to get into critical section will enter the CS.

    4.   Process Pr$_i$ will enter critical section after getting ($|p_i|$-1) "Go Ahead" message.

    5.   Calculate turnaround time and waiting time of the CS executing process

  }

  **else**

    Continue traversing;

**Aging section:** After process executes its CS, then the following actions are taken:

1. Increment the priority of all the requesting but not yet served processes in the GL by a random factor (a1)

$$node \rightarrow priority = node \rightarrow priority - a1$$

2. Promote them to upper levels, if there any changes in external priority.

*Priority of new aged process $=lowest\ priority\ among\ active\ processes\ of$*
*New level $+a$ positive random factor (a2)*

  **if** (node->aged == TRUE AND CS execution of that process has completed)

    Move back to its original level and place it at the end of the list.

  **else**

    Stay at that level.

  if any inactive process wishes to get into CS will get activated by changing request flag 0 to 1.

  } **while** (entire LL is not yet traversed==TRUE);

Move R to the next lower priority level by performing ***R=R->down***;

} **while**(R! = NULL);

**End**

$a1, a2 \in (0, 1)\ where\ a1 > a2$

The range of a1 and a2 maintains fairness.

**Illustration with an example**

We have created an adjacency list and according to external priority, we are assigning processes to each level. These levels are not static as the total number of processes increase the priority level may increase. Suppose the total number of participating processes are N. So the number of levels is Ceil (N/r), where $1 \leq r \leq N$, i.e., O (N). According to Fig. 3, processes A, B, E, H, I, K, M, and N are invoking (denoted by gray) for CS access. C, D, F, G, J, and L (denoted by black) remain silent for the time being. The priorities of processes are given in Fig. 3 and Table 1. In Table 1, the requesting processes are denoted by gray color and served processes are denoted by black. The process which is getting access to critical section is denoted by "+" and the aged process is denoted by "*". Process A is at root so A will send a request first to B and then to C. In turn, C will send a "go-ahead" message immediately to A because it is not requesting but B (1.8) has higher priority than A (1.9) so B will block A. LL contains A and B. GL contains A, B, E, H, I, K, M, N. Next, B sends request messages to A and C and eventually gets permission for CS access. Now LL contains process A and GL contains A, E, H, I, K, M, N. After B has completed accessing CS, A enters CS, as it has not been served. Other requesting processes that are present in levels below that level will be aged by a random factor (a1), here a1 = 0.3. Now new priority of $P_{ji}$, which means priority of jth process at the ith level will be $P_{ji}-0.3$. Now to keep fairness, if any process changes its priority level then its priority will be equal to the lowest priority of the active processes of new level + a2. As shown in Table 1 Process K and N had priority 4.2 and 4.1, respectively. After aging, their priorities will be 3.9 and 3.8, respectively. Minimum priority process at level 3 is H (3.2) and N has higher priority than K, so N will move up first. After changing priority level, the priority of N will be 3.2 + 0.01 (here a2 = 0.01) = 3.21 and priority of K will be 3.21 + 0.01 = 3.22. Now K has the lowest priority at level 3. After second CS access, there is no requesting process at level 1. So Highest priority process at level 2 that is E will get access. That is why after second CS the priority of E will not be increased. After second CS process M moves up to the priority level 3 and there is no requesting process, it will move up one more level and its priority will be 2.35. After completion of CS execution H, I, K, M, and N will move back to their original level because they had changed their levels. This process will stop when no process is requesting, i.e., GL is empty.

# 4 Performance Analysis

In this section, the proposed algorithm is evaluated from multiple perspectives. Issues considered for performance evaluation include correctness of the algorithm in terms of both progress condition and safety, message complexity, fairness, etc.

## 4.1 Message Complexity

For each critical section access, $2*(|Pi|-1)$ message exchanges are required. Pi is the number of processes present at level i.

**Table 1** Example of working of algorithm

| Processes | Priority | | | | | | | | | | |
|---|---|---|---|---|---|---|---|---|---|---|---|
| | Initial | 1st CS | | 2nd CS | | 3rd CS | 4th CS | 5th CS | 6th CS | 7th CS | 8th CS |
| A | 1.9 | 1.9 | 1.9$^+$ | **1.9** | **1.9** | **1.9** | **1.9** | **1.9** | **1.9** | **1.9** | **1.9** |
| B | 1.8$^+$ | **1.8** | **1.8** | **1.8** | **1.8** | **1.8** | **1.8** | **1.8** | **1.8** | **1.8** | **1.8** |
| E | 2.6 | 2.3 | 2.3 | 2.3$^+$ | 2.3$^+$ | **2.6** | **2.6** | **2.6** | **2.6** | **2.6** | **2.6** |
| H | 3.5 | 3.2 | 3.2 | 2.9* | 2.32 | 2.32 | 2.32$^+$ | **3.5** | **3.5** | **3.5** | **3.5** |
| I | 3.2 | 2.9* | 2.31 | 2.31 | 2.31 | 2.31$^+$ | **3.2** | **3.2** | **3.2** | **3.2** | **3.2** |
| K | 4.2 | 3.9* | 3.22 | 2.92* | 2.34 | 2.34 | 2.34 | 2.34 | 2.34$^+$ | **4.4** | **4.4** |
| M | 4.4 | 4.1 | 4.1 | 3.8* | 2.35 | 2.35 | 2.35 | 2.35 | 2.35 | 2.35$^+$ | **4.4** |
| N | 4.1 | 3.8* | 3.21 | 2.91* | 2.33 | 2.33 | 2.33 | 2.33$^+$ | **4.1** | **4.1** | **4.1** |

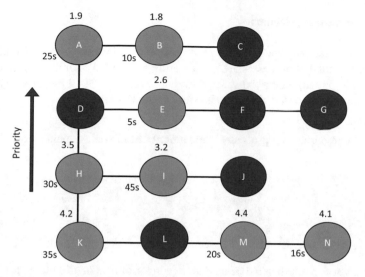

**Fig. 3** Pictorial representation of processes

## Expected Message Passing Complexity

$$= \frac{1}{n} \sum_{i=1}^{n} (2 * |Pi - 1|) \ldots \ldots \text{(i)} \text{Where } n \neq 0, n, Pi \in \mathbb{N}^{+}$$

$$= \frac{1}{n} (2P1 - 2 + 2P2 - 2 + 2P3 - 2 \ldots \ldots 2Pn - 2)$$

$$= \frac{1}{n} (2P1 + 2P2 + 2P3 + \ldots + 2Pn) - \frac{1}{n} * 2n$$

$$\frac{1}{n} \sum_{i=1}^{n} (2 * Pi) - 2 \ldots \ldots \ldots \text{(ii)}$$

### 4.1.1 Best-Case Scenario

The best case occurs when every priority level has just one node, i.e., $P_i = 1$ for all i, $1 \leq i \leq n$, as no message exchanges are required in such a situation. In the original Ricart–Agrawala algorithm, the best-case message complexity is $2(n-1)$.

### 4.1.2    Worst-Case Scenario

In the worst case, all processes have the same external priority. Thus, we infer that
P1 = N, i.e., total N processes are participating and all have same external priority.
So message passing complexity per critical section is 2(P1−1) = 2(N−1). In the
worst case, our proposed algorithm is equal to the RA algorithm.

**Delay due to aging**
If there are m requesting processes, the amount of time delay spent on aging would
be

$$= \sum_{i=0}^{m} (m - i)$$
$$= O\left(m^2\right).$$

If m is very small, then the delay due to aging is almost negligible, but for a good
distribution, a certain delay has been encountered.

**Lemma 1** *Proposed approach is starvation free.*

*Proof* The proposed algorithm is said to be starvation free, if every requesting process eventually is allowed access to its critical section. We prove this lemma by the
method of contradiction. We propose that

$Low - priority\ processes\ will\ starve\ using\ the\ proposed\ algorithm$    . . . hypothesis 1

Let A and B be two requesting processes, where A has the highest priority and B
has the lowest. Hence, A and B occupy the first and last subgroup of the request set
respectively. Let us assume that A is allowed access to its critical section and, as per
Proposition 1, process B never gets entry to its critical section.

According to the proposed algorithm, the priority of the all requesting and yet
to be served processes are increased by the same finite random value every time a
process is allowed access to its CS. Hence, eventually, the requesting processes move
to the upper level subgroups as their priority values exceed the external priority value
of their assigned subgroup after a finite interval. Also, once all the requesting nodes
have been served at the topmost priority level, the proposed algorithm forces it to
descend to the next level. Thus, eventually, B will be allowed to access its CS due to
this upward movement in the priority hierarchy after every access.

Hence, our initial hypothesis that "*low priority processes will starve using the
proposed algorithm*" is found to be false even for process B that has the lowest
priority. This proves that the converse of our initial hypothesis is correct. Thus, the
proposed algorithm is starvation free.

**Lemma 2** *Mutual exclusion holds correctness.*

*Proof* An algorithm is said to follow mutual exclusion if no two requesting processes enter the critical section simultaneously. In the proposed algorithm requests are served group wise. A process is said to enter its critical section if it has the highest priority among all requesting and yet to be served processes and can enter the critical section on receiving "go-ahead" replies from all other processes in the group. We evaluate two cases for the proof:

**Case 1:**
Let A, B, C, and D be the requesting processes with priorities $p_a$, $p_b$, $p_c$, and $p_d$ where $p_b > p_c > p_a > p_d$. The requests have arrived in the order as B, C, A, D. B at first sends a request message to all others in the group. Everyone else compares their priorities and sends a "go-ahead" to B refraining them from entering CS. B on receiving 3 "go-ahead" messages gets permission to enter CS and it does.

**Case 2:**
Let A, B, C, and D be the requesting processes with priorities $p_a$, $p_b$, $p_c$, and $p_d$ where $p_b > p_c > p_a > p_d$. The requests have arrived in the above order as well. At first, A sends a request message to all others in the group. Everyone else compares their priorities and sends a "go-ahead" to A except B and C, hence refraining A from entering CS. In the next round, B gets a "go-ahead" from all others and enters CS exclusively.

Generalizing this notion, if the level being served has k processes, then a process can enter CS if and only if its priority is highest among all currently requesting processes and receives a "go-ahead" reply from $k-1$ processes. Hence mutual exclusion is always satisfied.

**Lemma 3** *Progress condition is guaranteed*

*Proof* Progress condition for a mutual exclusion algorithm demands that one of the contending processes for the critical section will eventually be allowed to enter the CS, even when no single process gets majority votes.

From Lemma 2, we can say that the proposed algorithm selects a requesting process to enter its critical section if and only if it gets a "go-ahead" reply from all other processes in its group. The processes which are not interested in entering CS simply do not turn their request flag on and hence send a "go-ahead" directly. The other requesting processes compare their priority with the present priority and send a "go-ahead" accordingly. Thus at some point, the process with present highest priority among the requesting processes gets a "go-ahead" from the remaining processes and enters its critical section. Thus eventually, every requesting process becomes the process with the highest priority.

**Lemma 4** *Proposed approach is maintaining fairness*

*Proof* An algorithm is said to maintain fairness if it satisfies requests in the proper requested order along with priority constraints. We prove this lemma by the method of contradiction. We propose that

*Fairness will not be held for the proposed approach* ... hypothesis1

Let A and B be two requesting processes in the presently served level with priorities $p_a$ and $p_b$ with $p_a > p_b$. We shall assume that B gets access before A to its critical section. Let us also assume that B has requested before A. As per the proposed algorithm, B sends a request message to A, A compares its priority with B. On seeing its value is greater, by Lemma 2 it does not send a "go-ahead" message, hence B does not enter its critical section. After that, it is A's turn. A now repeats the algorithm and gets a "go-ahead" from B and enters its critical section, hence violating our initial assumption that B enters first.

This proves that the converse of our initial hypothesis that *"Fairness will not be held for the proposed approach"* is correct. Hence, the proposed algorithm maintains.

# 5   Results for Simulation

In order to assess message complexity, the Proposed Algorithm (PA) is compared with Ricart–Agrawal's algorithm (RA). In this chapter, the experimental setup has been documented in Table 2.

**Comparative Study between RA and the Proposed Approach**
We have done our experiment by taking multiple processes with the different priorities. We have done a comparative study based on the total number of valid message exchanges for all processes, average turnaround time, and average waiting time. In our dataset, we have assumed that there are 10 priority levels and each priority level

**Table 2**  Simulation Parameters

| Parameters | Value |
|---|---|
| Platform/Software | C language |
| Connection topology | Distributed complete subgraphs |
| Nodes in the graph | N |
| Edge length | 1 |
| Maximum out-degree | 3 |
| Maximum degree | 3 |
| Minimum degree | 1 |
| Priority of nodes in the graph | 0–N |
| Maximum number of candidates | N |
| Candidate select | Priority-level wise |
| Clock | System clock |
| Clock type | System clock |
| Execution time | O(N) |

**Table 3** Comparison between RA algorithm and PA

| Number of processes | Ricart–Agrawala algorithm(RA) | | | Proposed algorithm (PA) | | |
|---|---|---|---|---|---|---|
| | Average turn around time (TAT) | Average waiting time (WT) | Total message exchanges | Average turn around time (TAT) | Average waiting time (WT) | Total message exchanges |
| 100 | 1262 | 1237 | 19800 | 658.75 | 633.75 | 2180 |
| 200 | 2512 | 2487 | 79600 | 671.25 | 646.25 | 9160 |
| 400 | 5012 | 4987 | 319200 | 2018.94 | 1993.94 | 34145 |
| 600 | 7512 | 7487 | 718800 | 4704.95 | 4679.95 | 74224 |
| 800 | 10012 | 9987 | 1278400 | 7240.59 | 7215.59 | 129824 |
| 1000 | 12512 | 12487 | 1998000 | 9761.98 | 9736.98 | 201424 |

initially contains an equal number of processes with different internal priorities. Each process has burst time 25 ms and all processes have requested for access to the critical section. Our experimental data and results are given in Table 3. Average turnaround time and average waiting time were calculated in ms.

## 5.1 Message Complexity

It is observed from the Table 3 that when the number of the processes increases, total message complexity will also be increased. Total message exchanges will be much higher if we use RA algorithm where the proposed approach gives lesser number of message exchanges. In Fig. 4, it is observed that RA algorithm gives an exponential performance with respect to all the participating nodes, whereas the proposed algorithm gives linear performance observed under general cases. So, on an average, the proposed algorithm performs much better than the standard RA.

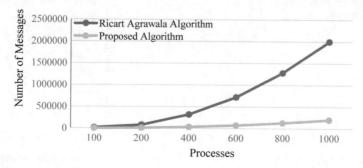

**Fig. 4** Message complexity for RA versus PA

## 5.2 Comparison of Turnaround Time (TAT) and Waiting Time (WT)

It has been observed from Table 3, Figs. 5 and 6 that the proposed approach produces better average turnaround time and the average waiting time than RA algorithm. A process has to wait less amount of time after requesting for access to critical section.

**Height of the structure**
The maximum number of priority levels is bounded by the number of participating nodes, i.e., it is n in the worst case. However, this is a rare scenario and only a theoretical possibility because the requesting nodes move to the upper levels to avoid starvation. Experimental observation suggests that in any case the number of message exchanges by the proposed algorithm shall never exceed RA algorithm. A linear bound (as obtained in the graph) can be obtained by an even distribution of

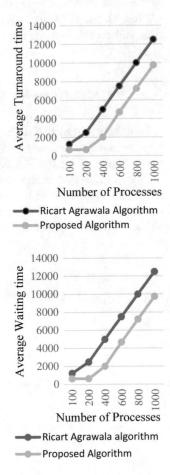

**Fig. 5** Average turnaround time for RA versus PA

**Fig. 6** Average waiting time for RA versus PA

the nodes into priority levels. For example, if there are 1000 participating nodes, distributing them into hundred priority levels each with a set of ten nodes gives a fair and symmetric distribution and helps us to obtain a linear time output in comparison to RA. Thus, the height of n/10 achieves a very good distribution.

## 6 Conclusions

The proposed algorithm contains one or more priority levels and one or number of processes is placed in each level. It is established in the text that the proposed algorithm maintains safety, liveness, and fairness. The theoretical analysis and experimental results presented in Sects. 4 and 5, respectively, establish that the message complexity and execution time for the proposed solution is better than the existing solutions compared. The number of message exchanges per critical section access is in the range of 0 to $2(N-1)$.

## References

1. Ricart, G., Agrawala, A.K.: An optimal algorithm for mutual exclusion in computer networks. Commun. ACM **24**(1), 9–17 (1981)
2. Lamport, L.: Time, clocks, and the ordering of events in a distributed system. Commun. ACM **21**(7), 558–565 (1978)
3. Lodha, S., Kshemkalyani, A.: A fair distributed mutual exclusion algorithm. IEEE Trans. Parallel Distrib. Syst. **11**(6), 537–549 (2000)
4. Kanrar, S., Chaki, N.: FAPP: a new fairness algorithm for priority process mutual exclusion in distributed systems, special issue on recent advances in network and parallel computing. Int. J. Netw. **5**(1), 11–18 (2010). ISSN 1796-2056
5. Raymond, Kerry.: A tree-based algorithm for distributed mutual exclusion. ACM Trans. Comput. Syst. **7**(1), 61–77 (1989)
6. Lejeune, J., Arantes, L., Sopena, J., Sens, P.: Service level agreement for distributed mutual exclusion in cloud computing. In: 12th IEEE/ACM International Conference on Cluster, Cloud and Grid Computing (CCGRID'12) (2012)
7. Lejeune, J., Arantes, L., Sopena, J., Sens, P.: A fair starvation-free prioritized mutual exclusion algorithm for distributed system. J. Parallel Distrib. Comput. (2015)
8. Swaroop, A., Singh, A.K.: A distributed group mutual exclusion algorithm for soft real-time systems. Proc. World Acad. Sci. Eng. Technol. **26**, 138–143 (2007)
9. Swaroop, A., Singh, A.K.: A token-based group mutual exclusion algorithm for cellular wireless networks, In: India Conference (INDICON-2009), pp. 1–4 (2009)
10. Housini, A., Trehel, M.: Distributed mutual exclusion token-permission based by prioritized groups. In Proceedings of the ACS/IEEE International Conference, pp. 253–259 (2001)
11. Maekawa, M.: A $\sqrt{N}$ algorithm for mutual exclusion in decentralized systems. ACM Trans. Comput. Syst. **3**(2), 145–159 (1985)
12. Atreya, R., Mittal, N., Peri, S.: A quorum-based group mutual exclusion algorithm for a distributed system with dynamic group set. IEEE Trans. Parallel Distrib. Syst. **18**(10), 1345–1360 (2007)
13. Kanrar, S., Choudhury, S., Chaki, N.: A link-failure resilient token based mutual exclusion algorithm for directed graph topology. In: Proceedings of the 7th International Symposium on Parallel and Distributed Computing (ISPDC) (2008)

14. Kanrar, S., Chaki, N., Chattopadhyay, S.: A new hybrid mutual exclusion algorithm in absence of majority consensus. In: Proceedings of the 2nd International Doctoral Symposium on Applied Computation and security System, ACSS (2015)
15. Singhal, M.: A heuristically-aided algorithm for mutual exclusion for distributed systems. IEEE Trans. Comput. **38**(5), 70–78 (1989)
16. Naimi, M., Thiare, O.: Distributed mutual exclusion based on causal ordering. J. Comput. Sci., 398–404 (2009). ISBN: 1549–3636
17. Suzuki, I., Kasami, T.: A distributed mutual exclusion algorithm. ACM Trans. Comput. Syst. (TOCS) **3**(4), 344–349 (1985)
18. Sayani, S., Das, S.: An energy efficient algorithm for distributed mutual exclusion in mobile ad-hoc networks. World Acad. Sci. Eng. Technol. **64**, 517–522 (2010)

# Part III
# Big Data and Analytics

# A Mid-Value Based Sorting

Najma Sultana, Smita Paira, Sourabh Chandra and Sk Safikul Alam

**Abstract** Proper ordering of data has always grabbed the attention of mankind. Researchers have always put forward their ideas for efficient sorting. A new method of sorting a large number of data in an array has been presented which is an iterative approach of sorting. The proposed method calculates an interval value from user input. Avoid repeated successive scan and repeated adjacent swap by comparing at the intervals. Promise to take less time to sort both small and large data than the existing sorting methods.

**Keywords** Sorting · Bubble sort · Insertion sort · Selection sort · Time complexity · Iterative · Online sorting

## 1 Introduction

Sorting is one of the key operations of data structure. It is the storage of data systematically either in ascending or descending order. Efficient sorting improves the performances of other data structure operations like searching, inserting, etc. [3]. Sorting methods are mainly categorized as traditional, divide and conquer and greedy methods depending on the flow of mechanism, computational complexity and space complexity [4].

Traditional sorting methods, namely, bubble sort, insertion sort and selection sort carry time complexity of $O(n^2)$ [2]. Bubble sort performs repeated adjacent swaps [7]. Insertion sort fetches an element and places it at the correct location in the array [5]. Selection sort repeatedly finds the smallest element and then swaps [8]. Traditional methods have better applications in educational organizations. Divide and conquer

N. Sultana · S. Chandra (✉) · S. S. Alam
Calcutta Institute of Technology, Uluberia, Howrah, India
e-mail: sourabh.chandra@gmail.com

S. Paira
IIEST Shibpur, Howrah, India

© Springer Nature Singapore Pte Ltd. 2018
S. K. Das and N. Chaki (eds.), *Algorithms and Applications*, Smart Innovation, Systems and Technologies 88, https://doi.org/10.1007/978-981-10-8102-6_5

approaches such as quick sort, merge sort, heap sort and radix sort of O(n log n) time complexity and greedy approach are way better in solving real-life problems [1].

Detailed analysis of the proposed algorithm based on mechanism, execution time, ease of implementation and brief comparison with the traditional sorting methods is done in the following sections.

## 2  Proposed Algorithm

The new algorithm calculates a block value from the number of input. Scanning from the left it jumps at a regular interval of block value and swaps if two data are out of order. Hence, this algorithm takes lesser time than the traditional algorithms as repeated successive scan and repeated adjacent swap get avoided. The below sub-sections introduce the pseudocode of the proposed algorithm followed by a detailed explanation of the steps of the algorithm using few elements in an array.

### 2.1  Pseudocode

Step 1: Initialize k to $(n/2)$ (n is even) or to $((n/2) + 1)$ (n is odd). Store k into m.
Step 2: Initialize k from Step 1.
Step 3: Initialize i to 0.
Step 4: If the value $(i + k)$ is less than n and if the value at position i is greater than the value at position $(i + k)$, swap them.
Step 5: Repeat Step 4 until i becomes $(n-k + 1)$.
Step 6: Decrement k by 1 and repeat Steps 3–5 until k becomes zero.
Step 7: Again initialize i to 0.
Step 8: If value at position i is greater than value at position $(i + 1)$, then swap them.
Step 9: Repeat Step 8 until i becomes m.
Step 10: Finally, print the sorted list.
Step 11: Exit.

### 2.2  Flow of Algorithm

Considering n element as input, block value(k) is calculated as n/2 for even value of n otherwise $((n/2) +1)$. While k is non-zero positive and decreasing by one, the first loop starts from 0 to $(n-k)$ and swaps if two data are out of order. Second loop starts from 0 to $(k-1)$ and swaps if out-of-order data are encountered and finally shows up the sorted list. The flow of mechanism of the proposed algorithm for an array size of 8 has been shown below:

Let the original array be

| 9 | 5 | 1 | 3 | 11 | 4 | 8 | 2 |
|---|---|---|---|----|---|---|---|
| 0 | 1 | 2 | 3 | 4 | 5 | 6 | 7 |

Step 1: K is calculated as n/2, i.e. 8/2 = 4.
Algorithm scans from left up to third position and compares two data at a regular interval of 4. If the two data are out of order, swap them, otherwise remain unchanged.

| 9 | 4 | 1 | 2 | 11 | 5 | 8 | 3 |
|---|---|---|---|----|---|---|---|
| 0 | 1 | 2 | 3 | 4 | 5 | 6 | 7 |

Step 2: k get decremented by 1, i.e. k = 3.
Algorithm scans from left up to fourth position and compares two data at a regular interval of 3. If the two data are out of order, swap them, otherwise remain unchanged.

| 2 | 4 | 1 | 9 | 3 | 5 | 8 | 11 |
|---|---|---|---|---|---|---|----|
| 0 | 1 | 2 | 3 | 4 | 5 | 6 | 7 |

Step 3: k get decremented by 1, i.e. K = 2.
Algorithm scans from left up to fifth position and compares two data at a regular interval of 2. If the two data are out of order, swap them, otherwise remain unchanged.

| 1 | 4 | 2 | 5 | 3 | 9 | 8 | 11 |
|---|---|---|---|---|---|---|----|
| 0 | 1 | 2 | 3 | 4 | 5 | 6 | 7 |

Step 4: K get decremented by 1, i.e. K = 1.
Algorithm scans from left up to sixth position and compares two data at a regular interval of 1. If the two data are out of order, swap them, otherwise remain unchanged.

| 1 | 2 | 4 | 3 | 5 | 8 | 9 | 11 |
|---|---|---|---|---|---|---|----|
| 0 | 1 | 2 | 3 | 4 | 5 | 6 | 7 |

Step 5: Now the algorithm scans from left up to Kth (i.e. fourth in this case) position and does adjacent swap if out of order encountered else remains unchanged.

| 1 | 2 | 3 | 4 | 5 | 8 | 9 | 11 |
|---|---|---|---|---|---|---|----|
| 0 | 1 | 2 | 3 | 4 | 5 | 6 | 7  |

Hence, the sorted list in ascending order is as follows:

| 1 | 2 | 3 | 4 | 5 | 8 | 9 | 11 |
|---|---|---|---|---|---|---|----|
| 0 | 1 | 2 | 3 | 4 | 5 | 6 | 7  |

## 3   Implementation of the Proposed Algorithm Using C

```c
#include<stdio.h>
#include<conio.h>
void sort(*b,p);
void main()
{
        int a[20],i,n;
        printf("Enter the number of elements");
        scanf("%d",&n);
        printf("\n Enter the elements\n");
        for(i=0;i<n;i++) //taking input from user
        {
                scanf("%d",&a[i]);
        }
        printf("\n Print the unsorted list\n");
        for (i=0;i<n;i++) //displaying the unsorted array
        {
                printf("%d",a[i]);
        }
        sort(a,n);
        printf("\n The sorted array is....\n");
        for(i=0;i<n;i++) //displaying the sorted array
        {
                printf("%d ",a[i]);
        }
getch();
} //end of main
```

```
void sort(*b,p)
{
        int i,k,m,p,temp;
        k=p/2;
        if(p%2!=0)
        {
                k=k+1;
        }
        m=k;
        while(k>0)
        {
                for(i=0;i<=p-k; i++)
                {
                        if (a[i+k]<p)
                        {
                                if(a[i]>a[i+k])
                                {
                                        temp=a[i];
                                        a[i]=a[i+k];
                                        a[i+k]=temp;
                                }
                        }
                } //End of For loop
                k=k-1;
        } //End of while loop
        for(i=0;i<m;i++)
        {
                if(a[i]>a[i+1])
                {
                        temp=a[i];
                        a[i]=a[i+1];
                        a[i+1]=temp;
                }
        }// End of For Loop
}//End of sort()
```

## 4   Time Complexity Analysis

Considering n number of elements in an array, k be the block size such that initially k = n/2.

The number of repetitions of step 6 in the above pseudocode depends on the successive values of k and the number of iterations required by Steps 4 and 5 as given below:

Number of comparisons required for first scanning

$$= \{(n/2) + 1\} + \{(n/2) + 2\} + \{(n/2) + 3\} + \ldots\ldots\ldots + \{(n/2) + (n/2)\}$$

$$= (n/2)^2 + 1 + 2 + 3 + \ldots\ldots\ldots + (n/2)$$

$$= (n/2)^2 + (n/4) \{(n/2) + 1\}$$

$$= \left(3n^2 + 2n\right)/8 \tag{1}$$

The second scan depends on Step 9 which in turns repeats for (n/2) number of times.

Hence, the numbers of comparisons required for second scanning

$$= (n/2) \tag{2}$$

Thus, the total time complexity $= \{(3n^2 + 2n)/8\} + (n/2)$ [from Eqs. (1) and (2)]

$$= \left(3n^2 + 6n\right)/8$$

## 5   Comparison Study

This section constitutes of a brief theoretical revision of traditional sorting algorithms like bubble sort, selection sort, insertion sort and a comparison study with the proposed algorithm, i.e. mid-value sorting.

Bubble sort, occasionally known as sinking sort, is the simplest way of sorting. It repeatedly compares adjacent items and swaps if required until the list is sorted [9]. It is easy to code. Unlike insertion sort, both bubble sort and selection sort exchange elements. Despite having same comparison number, selection sort has high performance over bubble sort as it requires less numbers of exchanges [4].

Insertion sort is effective one in doing online sorting [8] also outperform bubble sort by doing less swaps [6]. Both selection sort and insertion sort perform same comparisons but the latter is almost every time outperformed by first as it has O(n) swaps while the other one has O(n²) swaps. Also, large number of element shift in insertion sort is pricey in nature [11]. While the performance selection sort gets easily influenced by initial ordering of the data [10]. All these three algorithms perform well on

**Table 1** Execution time and number of comparisons in different sorting algorithms

| Sorting algorithm | Number of elements | No of comparisons required | Execution time (ms) |
|---|---|---|---|
| Mid-value sort | 100 | 3825 | 0.018 |
| | 200 | 15150 | 0.067 |
| | 500 | 94125 | 0.384 |
| | 1000 | 375750 | 1.666 |
| Bubble sort | 100 | 4950 | 0.05 |
| | 200 | 19900 | 0.143 |
| | 500 | 124750 | 0.849 |
| | 1000 | 499500 | 3.373 |
| Selection sort | 100 | 4950 | 0.028 |
| | 200 | 19900 | 0.072 |
| | 500 | 124750 | 0.5 |
| | 1000 | 499500 | 1.99 |
| Insertion sort | 100 | 4950 | 0.035 |
| | 200 | 19900 | 0.118 |
| | 500 | 124750 | 0.702 |
| | 1000 | 499500 | 3.216 |

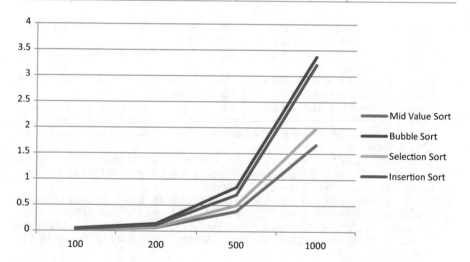

**Fig. 1** Graphical representation of time complexity of different sorting algorithms

small data [4]. Whereas the newly proposed algorithm is found to have fewer comparisons (i.e. $((3n^2 + 6n)/8))$ than the three and also has high performance on large data.

Unlike the existing methods, the newly proposed algorithm consists of two scanning. First scanning repeatedly compares at interval positions in the list until block value becomes zero from k. Second scanning swaps adjacent elements of the half of the list and produces the sorted list. Found to have less computational time to sort than the traditional sorting methods. The comparison table for execution time of different sorting methods along with number of comparisons and a graphical representation of time complexity is shown in Table 1 and Fig. 1, respectively.

# 6  Conclusion

Introduction of a new sorting method and a brief comparison study with the existing sorting methods has been done. The execution timetable followed by the graphical representation of time complexity of different sorting methods helped to prove the effectiveness of the new algorithm. It is found to have fewer comparisons and swaps to sort a large number of data. It is an initialization of more efficient future works. Inventions are nothing without real-life applications. Improvisation of this algorithm for real-life applications and comparable research works are assured in near future.

# References

1. Langsam, Y., Tenenbum, A.M.: Data Structures Using c and c ++, 2nd edn. Indian printing (Prentice Hall of India private limited), New Delhi-110001
2. Lipschutz, S.: Data Structure & Algorithm, 2nd edn. Schaum's Outlines Tata McGraw Hill, ISBN13: 9780070991309
3. Cormen, T.H., Leiserson, C.E., Rivest, R.L., Stein, C.: Introduction to Algorithms, 2nd edn. MIT Press and McGraw-Hill (2001). ISBN 0-262-03293-7. Problem 2-2, pp. 38
4. Knuth, D.: The Art of Computer Programming, vol. 3: Sorting and Searching, 3rd edn. Addison-Wesley (1997). ISBN 0-201- 89685-0. pp. 106–110 of section 5.2.2: Sorting by Exchanging
5. Ergül, O.: Guide to Programming and Algorithms Using R. Springer, London (2013). ISBN 978-1-4471-5327-6
6. Yu, P., Yang, Y., Gan, Y.: Experiment analysis on the bubble sort algorithm and its improved algorithms. In: Information Engineering and Applications. Lecture Notes in Electrical Engineering, vol. 154, pp. 95–102. Springer, London (2002)
7. Min, W.: Analysis on Bubble Sort algorithm optimization. In: 2010 International Forum on IEEE Information Technology and Applications (IFITA), Kunming, China (2010)
8. Yu, P., Yang, Y., Gan, Y.: Experimental study on the five sort algorithms. In: 2011 Second International Conference on IEEE Mechanic Automation and Control Engineering (MACE), China (2011)
9. www.sciencedirect.com/data_structure
10. www.answers.com/Q/What_are_advantages_and_disadvantages_of_selection_sort
11. www.answers.com/Q/What_are_advantages_and_disadvantages_of_insertion_sort

# Tracking of Miniature-Sized Objects in 3D Endoscopic Vision

Zeba Khanam and Jagdish Lal Raheja

**Abstract** The advent of 3D endoscope has revolutionized the field of industrial and medical inspection. It allows visual examination of inaccessible areas like underground pipes and human cavity. Miniature-sized objects like kidney stone and industrial waste products like slags can easily be monitored using 3D endoscope. In this paper, we present a technique to track small objects in 3D endoscopic vision using feature detectors. The proposed methodology uses the input of the operator to segment the target in order to extract reliable and stable features. Grow-cut algorithm is used for interactive segmentation to segment the object in one of the frames and later on, sparse correspondence is performed using SURF feature detectors. SURF feature detection based tracking algorithm is extended to track the object in the stereo endoscopic frames. The evaluation of the proposed technique is done by quantitatively analyzing its performance in two ex vivo environment and subjecting the target to various conditions like deformation, change in illumination, and scale and rotation transformation due to movement of endoscope.

**Keywords** Stereo endoscopic vision · Object tracking · Kidney stones
Feature detection

## 1 Introduction

Endoscope which started as a rudimentary device to examine hollow cavity has come long way. Advancements in the field of lens designing and fiber optics have made this device as one of the most powerful tools of the present day medicine and industry. The advantage of visual examination of hollow cavity without creation of cuts and large

Z. Khanam (✉)
Department of Computer Engineering, Aligarh Muslim University, Aligarh, India
e-mail: zeba.khanam@zhcet.ac.in

J. L. Raheja
CSIR-Central Electronics Engineering Research Institute, Pilani, India
e-mail: jagdish@ceeri.res.in

© Springer Nature Singapore Pte Ltd. 2018
S. K. Das and N. Chaki (eds.), *Algorithms and Applications*, Smart Innovation,
Systems and Technologies 88, https://doi.org/10.1007/978-981-10-8102-6_6

incisions has made it not only a diagnosis tool but also a crucial part of upcoming minimum invasive and robotic surgery.

The 2D display of the region of interest was the major huddle for application of endoscope in visualization of internal body parts. Loss of depth perception causes visual misperceptions. To compensate for this lack, 3D visualization using stereo vision has been employed. This technological innovation has accelerated the transition from open surgical methods to minimally invasive surgery. Surgical robots like da Vinci Surgical System uses 3D endoscope for visualization for complex neurosurgery, maxilla-facial and orthopedic surgery. Similarly, industry also comprises of environments which are inaccessible and demand remote visual inspections. Therefore, endoscopes have been deployed for rapid, nondestructive internal assessments of objects in industrial environment. In recent years, 3D endoscopes have been used to carry out safety checks in aerospace, pipelines, heavy plants, and manufacturing industry to name few.

In this paper, we elucidate an algorithm in which video stream from the stereo endoscopic camera is used to track miniature-sized objects. These objects because of their appearance and size are difficult for the operator to track using naked eyes. This computer vision framework will provide better assistance to doctors and inspectors. One such case where this technique will be useful is the peculiar cases of the ectopic pelvic kidney with ureteropelvic junction obstruction and presence of renal stones [17]. Tracking of dust particles in blast furnace for improving gas flow distribution [5].

In recent years, computer vision techniques have been used to computationally infer 3D tissue surface, soft tissue morphology, and surgical motion using 3D endoscope [20]. Current research focuses on addressing the issue of organ shift and soft tissue tracking [2, 7]. Several studies in past have tracked regions using methods based on optical flow, time of light, structured lightening, natural anatomical features, and fiducial markers [19]. However, direct application of tracking techniques to the endoscopic vision of MIS is not possible due to the non-Lambertian surface and contrastingly different visual appearances during surgery [18]. In [8], a probabilistic framework was proposed to track anisotropic features. Extended Kalman filter was designed to model the properties of affine invariant regions. However, this methodology explored soft tissue tracking and targeted the problem of free scale tissue deformation. Various techniques have been proposed to detect surgical instruments and track them during MIS. Earlier, external tracking systems were used which resulted in calibration errors. Later on, many works were produced without the use of external tracking system. These methodologies rely on discriminative region statistics between the target object and background [1]. However, no work targeting tracking of miniature-sized rigid objects was found in the literature. Therefore, in this work, inspiration was taken from the works trageting tracking of tissues and organs in 3D endoscopic vision where feature detection techniques were used.

Feature detectors based tracking have been used as some feature detectors are invariant to the global transformation. The extensions of Harris and Hessian corner detectors have been used to detect affine invariant region [14, 15]. Edge based region (EBR) [23] is one of the view invariant detectors used for localization and

**Fig. 1** A kidney stone

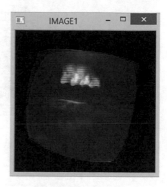

tracking. Maximally stable extremal regions [13] used watershed-based segmentation algorithm for tracking of affine transformations. MSER features are invariant to monotonic changes in illumination. They are able to successfully respond to MIS cardiac surface and differentiate superficial blood vessels or tissue bruising [21]. SIFT [12] and SURF [3] are able to successfully track objects using scale space approach of Laplacian-of-Gaussian (LoG) and Difference-of-Gaussian (DoG) operators. Figure 1 [11] shows the kidney stone. The irregular shape of stone does not allow detection of stone using edge or corner. Moreover, direct application of the feature detectors is not possible due to the constant change in the tracking scenes. Therefore, we have proposed a technique which uses an interactive segmentation algorithm and then applies view invariant detector to track features of the objects.

The paper has been organized into mainly four sections. In the first section, the problem targeted has been introduced along with brief description of related works. The second section of the paper explains the proposed framework in detail. The third section of the paper elucidates the experimental results obtained using the proposed technique. The fourth section summarizes the technique and discusses the result.

## 2 Proposed Method

The main aim of tracking is to provide further assistance to the operator who has limited view of the inspection scenes. A pair of stereo endoscopic camera is inserted through a small cavity for internal visualization. The interlaced output of the stereo camera allows the perception of depth which enhances dexterity of the operator. The proposed methods use 3D endoscopic vision as input and successfully track the stones in interlaced endoscopic vision. Figure 2 shows the flowchart depicting the steps used for tracking.

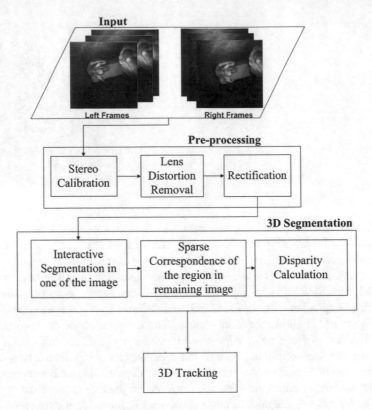

**Fig. 2** Flow chart of proposed technique

## 2.1 Experimental Setup

Awaiba NanEye 2C is used as a binocular camera. This stereo camera has a footprint of 2.2 × 1.0 × 1.7 mm. The dynamic illumination control is provided using fiber light source at the tip of the camera as seen in Fig. 3b. Figure 3a shows the entire assembly of NanEye 2C.

## 2.2 Preprocessing

The stereo endoscope generates the left and right images of the scenes which cannot be used directly for the interlace generation. Few mandatory steps are required as the preprocessing for correct 3D visualizations. As shown in Fig. 2, stereo calibration is the first step where intrinsic parameters and extrinsic parameters of the stereo camera are obtained. It also computes the geometrical relationship between left and right camera. Zhangs method [25] of stereo calibration using chessboard is applied.

**Fig. 3** **a** NanEye assembly **b** Camera head

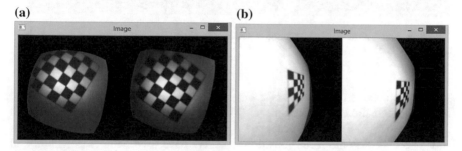

**Fig. 4** Side-by-side view of left and right images after application of rectification using **a** Bouguet algorithm **b** Hartley algorithm

Harris corner detector [9] is used to detect internal corners of the chessboard. The detected corner points serve as the reference point for stereo calibration. Miniature size of the lens causes the stereo camera to suffer from severe lens distortion. Radial lens distortion, which is mainly due to shape of the lens, is removed. The last step in preprocessing phase is rectification. It aligns left and right images into same plane in the world space. This ensures that respective coordinate of the pixel in other image lies in the same row. This reduces the computational time during sparse correspondence. Bouguet algorithm [4] was used, as it considers the calibration and distortion coefficients and gave acceptable results in comparison to Hartley algorithm [10] as shown in Fig. 4.

## 2.3 3D Segmentation

**Interactive Segmentation** The operator selects the target to be tracked. Therefore, seeded segmentation is applied to either left or right image. In our case, left image was selected for interactive segmentation. Figure 5a illustrates the background and foreground selection. The seeds are selected and indicated with a 5 cm filled circle around the point of selection.

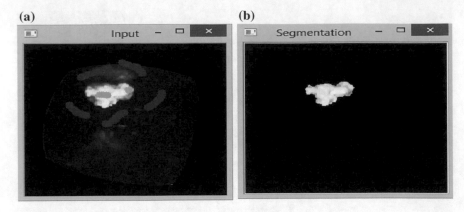

**Fig. 5 a** Foreground and background selection indicated by green and blue respectively **b** Segmented object

Segmentation is performed using the grow-cut algorithm [24] which is suitable for the segmentation of medical images [6]. It assumes each pixel of the image to be a triplet $A = (S, N, \delta)$ where S is a non-empty state set, N is the neighborhood system and $\delta : S^N \rightarrow S$ is the local transition rule. We have used Moores neighborhood system, where

$$N(p) = \left\{ q \in Z^n : \| p - q \|_\infty : = \max_{i=1,n} | p_i - q_i | = 1 \right\} \tag{1}$$

Each cell $S_p$ has three parameters associated with it.

1. Label L assigned to each cell classifies the cell as a background, foreground, or unclassified pixel.
2. Strength $\theta \in [0, 1]$ indicates the membership strength of the label. The pixel with the highest strength ($\theta = 1$) indicates convergence to foreground or background class.
3. Cell feature vector $\vec{C_p}$ indicates RGB vector of the pixels intensity.

The selection of background and foreground seeds is followed by initial assignment of each cell (pixel).

1. If pixel p is selected as a background seed
   $l_p = Background, \theta_p = 1, \vec{C_p} = RGB_p$
2. If pixel p is selected as a foreground seed
   $l_p = Foreground, \theta_p = 1, \vec{C_p} = RGB_p$
3. If pixel p is not selected
   $l_p = unassigned, \theta_p = 0, \vec{C_p} = RGB_p$

The object is segmented by assigning each pixel either label of background or foreground. Algorithm 1 illustrates the procedure for segmentation of the object.

**Algorithm 1** Segmentation

**for** $\forall q \in N(p)$ **do**
   A pixel p is attacked by its neighbor
   **if** $g(\| \overrightarrow{C_p} - \overrightarrow{C_q} \|_2 .\theta_q > \theta_p)$ **then**
      Pixel p is overpowered
      $\theta_p = Foreground$
      $Q_{(q \mid d)} = g(\| \overrightarrow{C_p} - \overrightarrow{C_q} \|_2 .\theta_q$
      $l_p = l_q$
   **end if**
**end for**

Figure 5b shows the final segmented kidney stone. The kidney stone with the fuzzy boundary and irregular shape can be segmented.

**Sparse Correspondence** After segmentation of the object in left frame, the object in right frame needs to be segmented. Due to rectification, the search space of the object is restricted to 1D. SURF [3] is used as the feature point detection technique. Features are detected in the segmented object in left frame and the horizontal region in the right frame as illustrated in Fig. 6.

FLANN [16] was used for matching the features. Figure 7a illustrates the result after application of FLANN. The stable matches with minimum Euclidean distance are retained and remaining outliers are rejected as seen in Fig. 7b.

**Disparity** The disparity is calculated as the average of the difference between stable features in the left and right image. If d is the disparity calculated, then the object is segmented in the right image using algorithm 2.

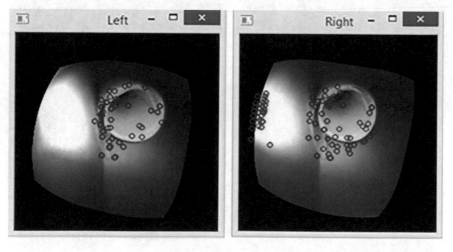

**Fig. 6** Features points in the object of the left image and corresponding horizontal patch in the right image

**Algorithm 2** Segmentation in Right Image

INITIALIZE each pixel of Right Image to also be a Cellular Automaton
**for** $\forall p \in RigthImage$ **do**
   $l_{(p)} = Background$
   $Q_{(p)} = 1$
   $\overrightarrow{C_p} = \overrightarrow{RGB}$
**end for**
**for** $\forall q \in LeftImage \&\& l_q = Foreground$ **do**
   Then q+d pixel in the Right Image is a Foreground Pixel
   $l_{(q+d)} = Foreground$
   $Q_{(q+d)} = 1$
   $\overrightarrow{C_{(q+d)}} = \overrightarrow{RGB}$
**end for**

The complexity of this algorithm is O(n), where n is the size of the image. In case the object is occluded, the algorithm segments the object in right frame, as a result of small baseline difference between the left and right sensor. Figure 8 shows interlace generated of the segmented kidney stone. This will allow the surgeon to perceive depth and estimate accurate size of the stone.

**(a)**                                      **(b)**

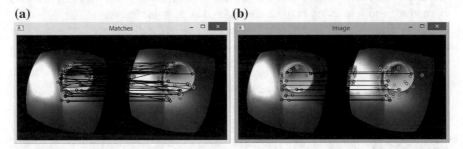

**Fig. 7** **a** All matches **b** matches with minimum distances

**Fig. 8** Segmented kidney stone

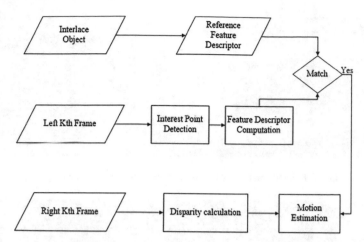

**Fig. 9** Tracking algorithm

## 2.4 *Tracking*

SURF-based tracking algorithm [22] is used for tracking objects in the monocular frames. This algorithm was extended for tracking objects in stereo frame. The object is detected in the left frame using SURF feature tracking. The object is tracked in the right frame using disparity algorithm mentioned above. Figure 9 illustrates the algorithm used for tracking. The tracked objects are marked on left and right frame and interlace of both the frames are generated for the surgeon to enable visualization in 3D.

## 3   Experimental Results

In order to validate our algorithm, two ex vivo simulating environments were created. The first environment is a customized matchbox with paper balls of different colors and varying sizes (few millimeters to 1 cm), shown in Fig. 10a. In the second environment, real kidney stones are placed in an enclosed box in Fig. 10b.

Performance of the technique proposed was evaluated under scale and rotation changes due to the movement of the endoscope, significant deformation of stones during surgery, and illumination changes. The stereo endoscope captures the left and right frames each of resolution 248 × 248 at 44 fps. Figure 11 illustrates the result obtained after subjecting the environment to the following four conditions:

1. Scale invariance
2. Rotation (±30°)
3. Illumination changes
4. Rigid deformation

**Fig. 10** **a** Environment inside matchbox **b** kidney stones based environment

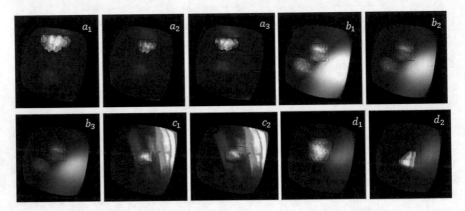

**Fig. 11** Tracking during $(a_1) - (a_3)$ scale changes $(b_1) - (b_3)$ illumination changes $(c_1) - (c_2)$ rotational changes $(d_1) - (d_2)$ deformation of stones

**Table 1** Mean and Standard Deviation of the error (in millimeter)

|             | Mean   | Standard deviation |
|-------------|--------|--------------------|
| Rotation    | 0.2441 | 0.0678             |
| Scaling     | 0.1612 | 0.0810             |
| Deformation | 0.2485 | 0.1346             |
| Illumination| 0.1004 | 0.1192             |

The error is calculated as the minimum Euclidean distance between stable feature point of actual region and detected region in the frame. Table 1 indicates the mean and standard deviation of the error calculated when tracking the stones in different conditions around 30 s (910 frames). It is evident from Table 1 that the technique performs better in the case of scale and illumination changes than rotation and deformation changes. However, results show that overall the technique is robust and successful in tracking miniature size objects.

# 4 Conclusion

In this paper, we have presented a robust technique for tacking miniature-sized objects in 3D endoscopic vision. The method presented is robust as demonstrated, and is able to track small objects under different environmental conditions. It detects the stable and reliable features for the object tracking using grow-cut algorithm. The extended SURF-based tracking algorithm is incorporated to track the miniature objects in the real-time stereo endoscopic vision. The results of experiment allow us to conclude that the proposed algorithms track stones successfully despite the scale and rotation changes due to the movement of endoscope. It is able to perform well during the illumination changes in the stereo endoscopic vision and deformation of the objects. Quantitative experiments on the simulated environment under various conditions help conclude that the technique is robust and reliable in tracking rigid objects of size varying from few millimeter to 5 cm. This work focuses on tracking a single object. Our future plan is to track multiple objects simultaneously. The experiments have been carried out by simulating the environment in the lab. However, in future, in vivo data will be used to assess the practical value of the algorithm accurately. The computational performance of technique will be improvised using optimal strategies for tracking. The work presented in this paper will serve as a computation tool for assistance to the surgeons and inspectors. We are hopeful that it will be a building block for the future tracking techniques for 3D endoscopic vision.

**Acknowledgements** This research work was financially supported by the CSIR-Network Project, "Advanced Instrumentation Solutions for Health Care and Agro-based Applications (ASHA)". The authors would like to acknowledge the Director, CSIR-Central Electronics Engineering Research Institute for his valuable guidance and continuous support. The authors would also extend their gratification to Birla Sarvajanik Hospital, Pilani for providing kidney stones for experiments.

# References

1. Allan, M., Ourselin, S., Thompson, S., Hawkes, D.J., Kelly, J., Stoyanov, D.: Toward detection and localization of instruments in minimally invasive surgery. IEEE Trans. Biomed. Eng. **60**(4), 1050–1058 (2013)
2. Baumhauer, M., Feuerstein, M., Meinzer, H.P., Rassweiler, J.: Navigation in endoscopic soft tissue surgery: perspectives and limitations. J. Endourol. **22**(4), 751–766 (2008)
3. Bay, H., Tuytelaars, T., Van Gool, L.: Surf: speeded up robust features. In: Computer Vision–ECCV 2006, pp. 404–417. Springer, (2006)
4. Bouguet, J.Y.: Camera calibration. Toolbox for Matlab (22 Nov 2010). Available from: http://www.vision.caltech.edu/bouguetj/calib_doc/index.html
5. Chen, Z., Jiang, Z., Gui, W., Yang, C.: A novel device for optical imaging of blast furnace burden surface: parallel low-light-loss backlight high-temperature industrial endoscope. IEEE Sens. J. **16**(17), 6703–6717 (2016)
6. Ghosh, P., Antani, S.K., Long, L.R., Thoma, G.R.: Unsupervised grow-cut: cellular automata-based medical image segmentation. In: 2011 First IEEE International Conference on Healthcare Informatics, Imaging and Systems Biology (HISB), pp. 40–47. IEEE, (2011)

7. Giannarou, S., Visentini-Scarzanella, M., Yang, G.Z.: Affine-invariant anisotropic detector for soft tissue tracking in minimally invasive surgery. In: IEEE International Symposium on Biomedical Imaging: From Nano to Macro, ISBI'09, pp. 1059–1062. IEEE, (2009)
8. Giannarou, S., Visentini-Scarzanella, M., Yang, G.Z.: Probabilistic tracking of affine-invariant anisotropic regions. IEEE Trans. Pattern Anal. Mach. Intell. 35(1), 130–143 (2013)
9. Harris, C., Stephens, M.: A combined corner and edge detector. In: Alvey Vision Conference, Citeseer, vol. 15, p. 50. (1988)
10. Hartley, R.I.: Theory and practice of projective rectification. Int. J. Comput. Vis. 35(2), 115–127 (1999)
11. Khanam, Z., Soni, P., Raheja, J.L., et al.: Development of 3D high definition endoscope system. In: Information Systems Design and Intelligent Applications, pp. 181–189. Springer, (2016)
12. Lowe, D.G.: Object recognition from local scale-invariant features. In: The Proceedings of the Seventh IEEE International Conference on Computer Vision, vol. 2, pp. 1150–1157. IEEE, (1999)
13. Matas, J., Chum, O., Urban, M., Pajdla, T.: Robust wide-baseline stereo from maximally stable extremal regions. Image Vis. Comput. 22(10), 761–767 (2004)
14. Mikolajczyk, K., Schmid, C.: Scale and affine invariant interest point detectors. Int. J. Comput. Vis. 60(1), 63–86 (2004)
15. Mikolajczyk, K., Tuytelaars, T., Schmid, C., Zisserman, A., Matas, J., Schaffalitzky, F., Kadir, T., Van Gool, L.: A comparison of affine region detectors. Int. J. Comput. Vis. 65(1–2), 43–72 (2005)
16. Muja, M., Lowe, D.G.: Fast approximate nearest neighbors with automatic algorithm configuration. VISAPP 1(2), 331–340 (2009)
17. Nayyar, R., Singh, P., Gupta, N.P.: Robot-assisted laparoscopic pyeloplasty with stone removal in an ectopic pelvic kidney. JSLS: J. Soc. Laparoendosc. Surg. 14(1), 130 (2010)
18. Puerto-Souza, G.A., Mariottini, G.L.: A fast and accurate feature-matching algorithm for minimally-invasive endoscopic images. IEEE Trans. Med. Imaging 32(7), 1201–1214 (2013)
19. Richa, R., Bó, A.P., Poignet, P.: Robust 3D visual tracking for robotic-assisted cardiac interventions. In: Medical Image Computing and Computer-Assisted Intervention–MICCAI, pp. 267–274. Springer, (2010)
20. Stoyanov, D.: Surgical vision. Ann. Biomed. Eng. 40(2), 332–345 (2012)
21. Stoyanov, D., Mylonas, G.P., Deligianni, F., Darzi, A., Yang, G.Z.: Soft-tissue motion tracking and structure estimation for robotic assisted mis procedures. In: Medical Image Computing and Computer-Assisted Intervention–MICCAI, pp. 139–146. Springer, (2005)
22. Ta, D.N., Chen, W.C., Gelfand, N., Pulli, K.: Surftrac: efficient tracking and continuous object recognition using local feature descriptors. In: IEEE Conference on Computer Vision and Pattern Recognition, CVPR 2009, pp. 2937–2944. IEEE, (2009)
23. Tuytelaars, T., Van Gool, L.: Matching widely separated views based on affine invariant regions. Int. J. Comput. Vis. 59(1), 61–85 (2004)
24. Vezhnevets, V., Konouchine, V.: Growcut: interactive multi-label ND image segmentation by cellular automata. In: Proceedings of the Graphicon, Citeseer, pp. 150–156. (2005)
25. Zhang, Z.: A flexible new technique for camera calibration. IEEE Trans. Pattern Anal. Mach. Intell. 22(11), 1330–1334 (2000)

# Author Index

© Springer Nature Singapore Pte Ltd. 2018
S. K. Das and N. Chaki (eds.), *Algorithms and Applications*, Smart Innovation,
Systems and Technologies 88, https://doi.org/10.1007/978-981-10-8102-6

Printed in the United States
By Bookmasters